Vision 2025:
How To Succeed in the
Global Chemistry Enterprise

ACS SYMPOSIUM SERIES **1157**

Vision 2025: How To Succeed in the Global Chemistry Enterprise

H. N. Cheng, Editor
Southern Regional Research Center
Agricultural Research Service
U.S. Department of Agriculture
New Orleans, Louisiana

Sadiq Shah, Editor
The University of Texas-Pan American
Edinburg, Texas

Marinda Li Wu, Editor
American Chemical Society
Washington, DC

American Chemical Society, Washington, DC

Distributed in print by Oxford University Press

Library of Congress Cataloging-in-Publication Data

Vision 2025 : how to succeed in the global chemistry enterprise / H.N. Cheng, editor, Southern Regional Research Center, Agricultural Research Service, U.S. Department of Agriculture, New Orleans, Louisiana, Sadiq Shah, editor, The University of Texas-Pan American, Edinburg, Texas, Marinda Li Wu, editor, American Chemical Society, Washington, DC.
   pages cm. -- (ACS symposium series ; 1157)
Includes bibliographical references and index.
ISBN 978-0-8412-2938-9 (alk. paper)
1. Chemical industry--Forecasting. I. Cheng, H. N. II. Shah, Sadiq. III. Wu, Marinda Li.
TP145.V57 2014
338.4'766--dc23

                        2014007159

Foreword

The ACS Symposium Series was first published in 1974 to provide a mechanism for publishing symposia quickly in book form. The purpose of the series is to publish timely, comprehensive books developed from the ACS sponsored symposia based on current scientific research. Occasionally, books are developed from symposia sponsored by other organizations when the topic is of keen interest to the chemistry audience.

Before agreeing to publish a book, the proposed table of contents is reviewed for appropriate and comprehensive coverage and for interest to the audience. Some papers may be excluded to better focus the book; others may be added to provide comprehensiveness. When appropriate, overview or introductory chapters are added. Drafts of chapters are peer-reviewed prior to final acceptance or rejection, and manuscripts are prepared in camera-ready format.

As a rule, only original research papers and original review papers are included in the volumes. Verbatim reproductions of previous published papers are not accepted.

ACS Books Department

Contents

Shared Experiences of
Successful Global Start-Ups and
Foreign Assignments

Indexes

Foreword

Thanks to all of you who have helped to support my vision as ACS President! I had contemplated writing a book if I got elected. In early 2012 soon after getting elected as ACS President-elect, I appointed a Presidential Task Force I decided to call "Vision 2025: Helping Members Thrive in the Global Chemistry Enterprise."

The goals for this Task Force were twofold: 1) Identify challenges and opportunities related to the global chemistry enterprise with respect to job growth, collaboration, education, and advocacy and 2) Provide recommendations to help members with jobs and to thrive in the global environment.

This hard working task force was ably co-chaired by Dr. H.N. Cheng, who led the working group focused on "Globalization Opportunities," and by Dr. Sadiq Shah, who led the working group on "Jobs and Advocacy."

As described in this book, we observed a number of trends affecting our members and others in the global chemistry enterprise. My Presidential Task Force studied both the challenges and the opportunities facing the global chemistry enterprise in order to help our members thrive and succeed.

In order to get more diverse perspectives, I invited eleven presidents of chemical societies from around the world to our 2013 spring national ACS meeting in New Orleans. These presidents represented many sister societies in Europe, Asia, Africa, and both North and South America. I also invited thought leaders from across the USA representing academia, industry, government, and small business.

Together, we discussed common challenges and opportunities facing the chemical industry and profession at Global Collaboration Roundtable discussions. Diverse perspectives, ideas, and experiences were shared by these invited speakers at the presidential symposium on "Globalization Opportunities." In order to help members beyond those able to attend this presidential symposium at the national ACS meeting, I invited each speaker to contribute a chapter toward this book to make the information accessible to both members and non-members of the global chemistry enterprise.

Thus, Drs. Cheng, Shah and I are delighted to be able to share the collective wisdom and ideas from a broad professional spectrum of our global chemistry enterprise. We hope that chemistry communities around the world will find this book informative, thought provoking, and a catalyst for action and partnering for progress and prosperity.

I want to express my sincere thanks to my co-editors and co-chairs of my Presidential Task Force, Dr. H.N. Cheng and Dr. Sadiq Shah, for making this book possible. We have enjoyed working together as ACS colleagues on various projects for many years. Special thanks are also due to Dr. Robert Rich, Director of Strategy Development for ACS, who ably supported my Presidential Task Force as the ACS staff liaison.

Thanks are also due to the many members of my Presidential Task Force, the presidents of the chemical societies represented in this book, and the speakers from my Presidential Symposium at the 2013 national ACS meeting on April 8-9, 2013 in New Orleans

I also want to thank my amazing daughter Lori, who married her talented Stanford classmate Evan, and my awesome son Will, who incidentally designed the Partners for Progress and Prosperity logo on the cover of this book. They inspire me to do what I can to help ensure we leave the world a better place for future generations.

Last but not least, I wish to thank two other very special people who have always supported my various endeavors in life—my dear mother Tsun Hwei Li, who is now 93 years old, and my wonderful husband of over 36 years, Norm.

As I have shared with our chemistry colleagues worldwide, we must continue to "Partner for Progress and Prosperity!" Please see my article in C&EN, Jan. 7, 2013, pg. 2 for more details.

Marinda Li Wu
2013 President, American Chemical Society

Preface

This book was developed from the American Chemical Society (ACS) Presidential Symposium on "Vision 2025: How to Succeed in the Global Chemistry Enterprise", held at the 245th National Meeting of the ACS in New Orleans in April 2013. The symposium speakers (and authors of the book chapters) were top leaders of the chemistry enterprise, including Presidents of international chemical societies, corporate executives, academic thought leaders, federal science agency director, and successful entrepreneurs. The purpose was to provide their collective perspectives on the global chemistry enterprise and share their experiences and ideas in order to benefit chemistry professionals and students in the future. Some of the topics covered included current activities of international chemical societies, successful global collaborative efforts, ideas on further cooperative and educational opportunities, and examples of recent successful research or entrepreneurial efforts.

A total of 22 chapters are included in this book with contributions from almost all symposium speakers. For convenience, they are divided into three sections: 1) Perspectives from U.S. leaders, 2) Perspectives from international leaders, and 3) Successful global start-ups, collaborations, and overseas assignments.

The final report and recommendations of the ACS Presidential Task Force on Vision 2025 (mentioned in the Foreword) are included as Chapter 22. This book should be useful to a wide range of audiences, from professors and senior corporate executives, to working scientists and students. They may be especially helpful to people seeking to foster international collaborations, to publicize chemistry to the public, to promote chemistry education, or to start new businesses. The younger scientists and students may take note of the trends and the changes in the global chemistry enterprise described in these chapters and manage their careers accordingly.

We appreciate the efforts of the authors who took time to prepare their manuscripts and our many reviewers for their cooperation during the peer review process. We thank the members of the ACS Presidential Task Force for their help in the past two years (Jens Breffke, Susan B. Butts, James Chao, Mukund S. Chorghade, Pat N. Confalone, Peter K. Dorhout, Dan Eustace, John Gavenonis, Jennifer S. Laurence, Zafra Lerman, Cynthia A. Maryanoff, Connie J. Murphy, Attila E. Pavlath, Dorothy J. Phillips, Al Ribes, Sonja Strah-Pleynet, Shaomeng Wang, Joel I. Shulman, Sharon Vercellotti, and Zi-Ling (Ben) Xue). Certainly we should also thank the ACS staff members, who ably assisted in the various tasks related to the Presidential symposium and Presidential Task Force, particularly Robert Rich, Brad Miller, Frank Walworth, and Alicia Harris. Encouragement and continuing support from Madeleine Jacobs and Denise Creech are also much appreciated.

H. N. Cheng
Southern Regional Research Center
Agricultural Research Service
U.S. Department of Agriculture
1100 Robert E. Lee Blvd.
New Orleans, Louisiana 70124, United States

Sadiq Shah
The University of Texas-Pan American
1201 W. University Drive,
Edinburg, Texas 78539, United States

Marinda Li Wu
American Chemical Society
1155 Sixteenth Street, NW
Washington, DC 20036, United States

Global Opportunities from U.S. Perspectives

Chapter 1

Partners for Progress and Prosperity in the Global Chemistry Enterprise

H. N. Cheng,*,[1] Sadiq Shah,[2] and Marinda Li Wu[3]

[1]Southern Regional Research Center,
Agricultural Research Service,
U.S. Department of Agriculture, 1100 Robert E. Lee Blvd.,
New Orleans, Louisiana 71024, United States
[2]The University of Texas-Pan American, 1201 W. University Drive,
Edinburg, Texas 78539, United States
[3]2013 President, American Chemical Society, 1155 Sixteenth Street NW,
Washington, DC 20036, United States
*E-mail: hn.cheng@ars.usda.gov

In the past several years, there have been many changes facing the global chemistry enterprise. Whereas the overall chemistry enterprise appears to be strong and the chemical industry is still a major contributor to GDP, many chemistry-based products have been commoditized, and chemical employment has been uneven, stronger in some countries and weaker in others. There is also persistent negative publicity towards chemicals and especially the chemical industry. To address such issues, ACS President Marinda Li Wu appointed a Presidential Task Force in 2012 to study both the challenges and opportunities. After much analysis and discussion, the Task Force developed seven recommendations dealing with jobs, advocacy, and globalization opportunities. Subsequently, the Task Force worked with 27 stakeholder national ACS committees and divisions to discuss implementation of these recommendations. Beneficial interactions have also been initiated with many sister chemical societies around the world as well as with our ACS international chapters. Already, several concrete steps towards implementation have been taken with respect to these seven recommendations. Our Task Force recommendations have been shared and well received by chemistry communities in academia, industry, and government both in the USA and worldwide. Working together, all of us can substantially help ACS members as well as other chemistry professionals thrive in the global chemistry enterprise. We are truly "Partners for Progress and Prosperity" where we can benefit by working together on common issues in order to transform challenges into opportunities.

Introduction

Chemistry is a central science, and chemistry-based products and services have a major impact on employment, trade and economic growth worldwide (*1*). Despite some ups and downs (*2*), chemical output is still a major contributor of GDP (*3*). However, several changes are taking place in the global chemistry enterprise. Many chemistry-based products have been commoditized (*4*). Production of chemicals has been shifting from highly industrialized countries to developing countries (*5*). In the U.S., since the 2008-2009 recession, chemistry practitioners have been hard hit by layoffs in the chemical and pharmaceutical industries (*6*) resulting in unemployment and underemployment especially in many industrial sectors (*7*). At the same time, budget deficits and debates on government finances constrain U.S. investment in science and engineering (*8*). In addition, there remains a persistent negative perception among the general public regarding chemicals in general and towards the chemical industry in particular (*9*).

The American Chemical Society (ACS) is a member-based professional organization and is concerned about the impact of these changes on its members. To address these challenges, Dr. Marinda Li Wu, ACS 2012 President-Elect (and 2013 President), appointed a Presidential Task Force entitled "Vision 2025: Helping ACS Members Thrive in the Global Chemistry Enterprise" in 2012 (*10*). The Task Force goals included identifying challenges and opportunities, and helping members to seek jobs and manage careers. The Task Force worked hard to study the issues, provide recommendations, and proceeded to implement them with the help of ACS governance and staff. This chapter provides an overview of the key activities of the Task Force. The final report of the Task Force is included as Chapter 22 in this book (*11*).

The Task Force was divided into two working groups: *Globalization Opportunities* chaired by Dr. H.N. Cheng, and *Jobs & Advocacy* chaired by Dr. Sadiq Shah. ACS President-elect Wu led the Task Force along with the two working group chairs, and Dr. Robert Rich (ACS Director, Strategy Development) provided valuable staff support. The Task Force appreciates the input and contributions of many members of ACS governance and staff and looks forward to continuing to partner as the recommendations are implemented. Since several recommendations have international implications, we intend to support ACS in future collaborations with our sister chemical societies and others to benefit the global chemistry enterprise.

Results and Discussion

Work Process

The two working groups held face-to-face meetings at both national ACS meetings and monthly conference calls in 2012 in addition to specific, targeted meetings and conference calls in 2013. They assessed the current landscape of the global chemistry enterprise; identified gaps, threats and opportunities; and ensured that any ideas recommended did not duplicate existing ACS efforts but instead would strengthen current efforts and leverage synergies that exist. The following approach was used:

1. Brainstormed strengths, weaknesses, opportunities, and threats (SWOT analysis).
2. Discussed related external trends and leading indicators.
3. Considered related existing and planned ACS activities.
4. Developed possible recommendations for new and enhanced offerings.
5. Consulted with key committees and other stakeholders on possible recommendations and refinements.
6. Presented draft recommendations to ACS Board of Directors at end of 2012 in a Task Force Report (*11*).
7. Revised recommendations and shared broadly to encourage implementation in 2013 and beyond.

Environmental Scan

Among the first tasks was a situation analysis. The following key trends and challenges were discussed:

■ The chemistry enterprise continues to be globalized.. Chemical products, R&D, manufacturing, and services (as well as associated jobs and capital) increasingly move across national boundaries (*12, 13*).

■ GDP growth is relatively low in the U.S., United Kingdom, Germany, and Japan. It is high in several developing countries (*14*).

■ Many chemical products have become commodities. The specialty chemical industry is especially affected by the commoditization, such that competition has increased and profit has fallen (*4, 15*).

■ Boundaries between chemistry and other sciences are becoming blurred as research increases at the interfaces. Today's jobs are crossing traditional disciplinary boundaries, and inter- and multi-disciplinary content in degree programs is increasing to support industry's needs (*16, 17*).

■ According to the U.S. Bureau of Labor Statistics and National Science Foundation (NSF) data, chemical jobs have been decreasing in the U.S. during the past 20 years and will continue to decline in the near future (*18*).

■ In 2012 the average unemployment rate in the country was 7.9%, and the unemployment rate for ACS member chemists is now 4.2% (a decline from the all-time peak of 4.6% in 2011 but still very high by historical standards) (*19*).

■ Unemployment among new graduates is more severe, averaging 13.3%, compared with 4.6% overall in chemistry in 2011 (*20, 21*).

■ Despite high unemployment for new chemistry graduates, the number of fresh M.S. and Ph.D. graduates has increased (*22, 23*).

■ Students are concerned about their future in chemistry and are experiencing difficulty navigating the job market. Graduate students, in particular, are concerned about the lack of jobs in industry and available faculty positions. Postdoctoral studies are often used to land entry-level positions within the private sector (*24*).

■ Students recognize the need to improve skills related to job search. Students and postdocs are looking for opportunities to network among academic colleagues and those in industry who can provide jobs (*25*).

■ Students and faculty note the need for better preparation for careers outside of graduate academic institutions, with which many faculty members are unfamiliar (*26*).

■ The U.S. is still the envy of the world as far as graduate education is concerned (*27, 28*).

■ U.S. academic institutions of higher education are increasingly building partnerships with universities in other countries for education and research. This provides a competitive advantage and global opportunities for those U.S. graduates who have had an international exposure as part of their education (*29*).

■ Unprecedented budget deficits and demands on government finances at all levels of U.S. government (and within the European Union) constrain its investment in science and engineering (*30*).

■ Continued world population growth results in the following global challenges for chemistry: affordable medicines and medical care, depletion of earth's resources, rising energy costs, air and water pollution, and ample food supply to meet worldwide demand.

■ The general public continues to have a negative perception about chemicals and the chemical industry.

Recommendations

In view of the trends and the challenges, key questions the Task Force addressed were: What should ACS do? What should a chemist do? Where will the jobs be in the future? How does globalization affect us? How can we transform challenges into opportunities?

The Task Force studied these issues and received very useful input from many ACS leaders, volunteers and staff. After considerable thought, analyses, and discussions, the following recommendations were formulated. These recommendations are consistent with the ACS Strategic Plan and built upon many ongoing programs at ACS:

1. Better educate ACS members about the critical elements necessary for success in a broad spectrum of career paths.
2. Strengthen ACS efforts to support entrepreneurship.
3. Engage and equip members with enhanced advocacy tools and training so that they can proactively contact their legislators to improve the business climate and aid jobs creation.
4. Explore with U.S. and global stakeholders the supply and demand of chemists/jobs to bring them to a better equilibrium.
5. Collaborate with others, including chemical societies around the world regarding public communication, education, advocacy, chemical employment, and other topics of mutual interest.
6. Provide information, resources, advice, and assistance to ACS members interested in global job opportunities.
7. Expand ACS support for chemists and chemistry communities worldwide.

Update on Follow-Up Activities

In April 2013 at the national ACS meeting, the Task Force findings and pertinent recommendations were shared with 27 stakeholder committees and divisions. During the summer of 2013, communication and discussions continued with the chairs of all these stakeholder groups on how they can best support implementation of the recommendations. Highlighted below are recommendations and some of the completed and ongoing activities as of October 2013. Please note the following acronyms: CA = Corporation Associates, CCPA = Committee on Chemistry and Public Affairs, C&EN = Chemical & Engineering News, CEPA = Committee on Economic and Professional Affairs, ComSci = Committee on Science, CPRC = Committee on Public Relations and Communications, CPT = Committee on Professional Training, GEAB = Graduate Education Advisory Board, IAC = International Activities Committee, I&EC = Industrial & Engineering Chemistry Division, MAC = Membership Affairs Committee, OIA = ACS Office of International Activities, OPA = ACS Office of Public Affairs, SCC = Senior Chemists Committee, SOCED = Society Committee on Chemical Education, YCC = Younger Chemists Committee.

(Recommendation #1: Broad Spectrum of Career Paths)

1. The Task Force has looked at a wide range of possible careers that someone trained in chemistry can consider, including teaching, research, product and process development, testing and analysis, consulting, sales, marketing, regulatory, technical service, patent, government policy, journalism, business development, project management, science writing, small business, venture capitalism, and more. At the ACS National Meeting in Indianapolis in September 2013, the Task Force organized a Presidential Symposium on Career Advancement Opportunities with speakers representing a wide variety of careers including as keynote speaker, Dr. John Lechleiter, the CEO of Eli Lilly. The ACS Career

Management and Development Department also has a tremendous collection of useful information and resources (*31*) for members interested in jobs and career development. Also see President Wu's ACS Comment " Looking for a Job? Check Out These Tools for Chemists" in *Chem. Eng. News* (*32*).

(Recommendation #2: Entrepreneurship)

2. The Task Force also organized a Presidential Symposium on Innovation and Entrepreneurship in Indianapolis featuring entrepreneurs from academia, small business and even a graduate student. Speakers shared personal stories and tips on how they became entrepreneurs. An update on the ACS Entrepreneurial Initiative was also presented by Dr. David Harwell, the staff liaison for this initiative.

(Recommendation #3: Enhanced Advocacy Tools and Training)

3. At Indianapolis, CCPA along with CEPA, CPRC, I&EC, SOCED, YCC and SCC all helped cosponsor a Presidential Advocacy Training Workshop called "React with Congress: Become a Chemistry Advocate." President Wu also wrote an ACS Comment in *Chem. Eng. News* (*33*) on "Time to Partner and Speak Up For Science" where more resources and power tools were given for use in advocacy. More information is also available on the ACS website (*34*).

(Recommendation #4: Supply and Demand of Chemists/Jobs)

4. Because several committees expressed interest in this important issue, President Wu formed a new Task Force to further study this issue for the U.S. CEPA agreed to take the lead with representatives from CA, CPT, ComSci, GEAB, CCPA, and YCC. This new Supply/Demand Task Force held its first meeting at the national ACS meeting in Indianapolis and plans to report its findings in 2014.

(Recommendation #5: Collaboration with Global Chemical Societies)

5. At the April 2013 ACS National Meeting in New Orleans, the Task Force organized a Global Opportunities Symposium. This symposium included eleven other presidents of chemical societies representing Europe, Asia, Africa and the Americas invited by President Wu to share their perspectives. Additional speakers included thought leaders from U.S. business, academia, and government. Other invited speakers shared their personal stories and experiences with global start-ups and successful overseas assignments.

6. In New Orleans, the Task Force also hosted a Global Collaboration Roundtable discussion with the 12 presidents of chemical societies.

This resulted in agreements to exchange contacts worldwide to improve public communication as well as generating the idea of producing a YouTube video for the general public demonstrating benefits that chemistry brings to society. Through the efforts of C&EN, OPA, IAO, and the ACS President's Office, the first ACS Global YouTube video contest was successfully launched in the summer of 2013, and a high school teacher won the contest with her video on "What to do if your dog gets skunked."

7. In Indianapolis, President Wu and ACS International Activities (IAC and IAO) hosted the first ACS International Chapter Summit on September 12-13, 2013. All six international chapters (Saudi Arabia, Hong Kong, Hungary, Shanghai, Thailand, and Romania) sent representatives to the meeting. Also present were IAC representatives and ACS staff. The agenda included information on member recruitment and incentives (presented by Dr. Wayne Jones, MAC Chair), best practices and exchanges, examples of successful programs such as Science Cafes and Festival de Química, strategic planning, annual reports, and brief overviews of three topics: Planning Successful Events, Engaging and Motivating Volunteers, and Engaging Colleagues in Dialogue.

(Recommendation # 6: Resources & Assistance to ACS Members on Global Job Opportunities)

8. The International Employment Initiative (IEI) was launched successfully in New Orleans for the very first time and offered again in Indianapolis at the two national ACS meetings in 2013. It surpassed our expectations and enabled international employers to connect with job seekers via our ACS Virtual Career Fair and/or in person. IEI has been met with great interest by international employers both from industry and academia and should continue to grow as awareness spreads across the global chemistry enterprise.

9. The Task Force enthusiastically endorsed the new ACS International Center, which is being operated by ACS International Activities. The International Center website contains comprehensive information on educational opportunities and international work experiences for the benefit of ACS members and potential members. Visit www.acs.org/ic.

(Recommendation #7: Expanded ACS Support for Chemists and Chemistry Communities Worldwide)

10. In Indianapolis, in celebration of its 20th anniversary, the Committee on Minority Affairs organized a Presidential Symposium on "The Impact of Diversity and Inclusion" with speakers representing all of the underrepresented groups from the Diversity and Inclusivity Advisory Board.

11. ACS International Activities continues their considerable international engagement activities (*35–37*). As part of their work, they have presented awards to recognize outstanding achievements in the international arena. For example, they presented a ChemLuminary Award for outstanding international engagement at the national ACS meeting in Indianapolis and poster awards to graduate students at international conferences in 2013. This has been an area that President Wu has long supported and encouraged—that is, giving more recognition and raising awareness of the outstanding chemistry and collaborations in the global chemistry enterprise.

Conclusions

Over the past 20 years, the number of U.S. jobs in the chemical sciences has steadily decreased. Many chemistry-based products have become commodities, and the chemistry enterprise has become more global. Layoffs and limited hiring from the private sector coupled with budget deficits have created tremendous challenges in the U.S. chemistry job market. The recommendations provided by the Vision 2025 Presidential Task Force will hopefully help mitigate some of the challenges, particularly with respect to jobs, advocacy, and globalization opportunities. It will be useful to continue discussions with all stakeholders in order to bring a better equilibrium between supply and demand for chemistry professionals. Job expansion can be sought in global, multidisciplinary, non-research, and non-traditional areas (Table 1). A major emphasis should be placed on providing information and assistance to prospective job seekers on diverse career possibilities, global opportunities, entrepreneurship, and career management strategies and skills. For details, please consult Section VI Subsection 5 (Jobs and the Future) in the Task Force Report in Chapter 22.

For the sake of the global chemistry enterprise and its practitioners, it is increasingly important for ACS to collaborate, to take advantage of the opportunities that globalization offers, and to address the challenges it creates. The Task Force appreciates the cooperation and the friendship of sister chemical societies. We hope to continue our collaborations, particularly in communication of chemistry's vital role to the public and policymakers, educational exchanges, as well as joint meetings and projects.

President Wu has been sharing the Task Force recommendations with local ACS sections, sister chemical societies, universities, corporations, government research labs, institutions and others at conferences both domestic and overseas. Her presidential message to audiences worldwide continues to be "Let's partner for progress and prosperity (*38, 39*)!"

Table 1. Types of jobs available to people with chemistry training (left column), possibilities of inter- or multi-disciplinary work (middle column), and global job opportunities (right column)

Chemistry/Science/U.S.	Multidisciplinarity	Global opportunities
Chemistry jobs: teaching, research, product R&D, engineering, analytical, testing, consulting	- chemistry - biochemistry - chem engineering	*Academe (international):* teaching, research, exchange program, collaboration, equipment use, postdoc
Chemistry-based: sales, manufacturing, marketing, regulatory, technical service, quality control	- biomed - pharma - materials - energy - nanotech - environment - ecology - regulatory	*Industry (overseas):* collaboration, foreign assignments, global teams, clinical testing, sales & manufacturing support
Science-related: patent work, government policy, grant officer, science journalism, business management, science policy, association management	- biology - computer science - food, agriculture - atmospheric science - geology - physics	*Small businesses:* start-up, import-export, contract synthesis, contract manufacturing, contract testing and analysis
Science-inspired: wall street, story writing, venture capitalist	- non-traditional (e.g., science related jobs)	*Others:* Foreign service (science attachés), teaching of English, editing of manuscripts, translation service

Acknowledgments

We thank the members of the Presidential Task Force on "Vision 2025: How to Thrive in the Global Chemistry Enterprise" for their commitment and hard work Alphabetically, it consists of Jens Breffke, Susan B. Butts, James Chao, H. N. Cheng, Pat N. Confalone, Mukund S. Chorghade, Peter K. Dorhout, Dan Eustace, John Gavenonis, Jennifer S. Laurence, Zafra Lerman, Cynthia A. Maryanoff, Connie J. Murphy, Attila E. Pavlath, Dorothy J Phillips, Al Ribes, Sadiq Shah, Joel I. Shulman, Sonja Strah-Pleynet, Sharon V Vercellotti, Shaomeng Wang, Marinda Li Wu, and Zi-Ling (Ben) Xue. Bob Rich serves ably as the ACS staff liaison. Thanks are also due to many ACS leaders, volunteers, and staff who generously shared their thoughts, time, and energy with us. Without the collective input, this work would not have been possible.

References

1. Global Chemicals Outlook, United Nations Environment Programme, 2012. http://www.unep.org/pdf/GCO_Synthesis%20Report_CBDTIE_UNEP_September5_2012.pdf.
2. Business Department. Lackluster year for chemical output. *Chem. Eng. News* **2013**, *91* (26), 41.
3. Guide to the Business of Chemistry – 2013. American Chemistry Council. www.americanchemistry.com/business-of-chemistry-summary.
4. Mullin, R. Lab-to-market connection. *Chem. Eng. News* **2008**, *86* (16), 15–23.
5. Thayer, A. M. Sourcing beyond China and India. *Chem. Eng. News* **2012**, *90* (5), 20–26.
6. Hanson, D. J. Starting salaries. *Chem. Eng. News* **2011**, *89* (11), 49–53.
7. Mullin, R. The future is now. *Chem. Eng. News* **2013**, *91* (49), 12–17.
8. Widener, A. Sequestration, shutdown dominated headlines. *Chem. Eng. News* **2013**, *91* (51), 26.
9. Raber, L. ACS takes public's pulse. *Chem. Eng. News* **2000**, *78* (41), 60–61.
10. Wu, M. L. Helping members thrive in the global chemistry enterprise. *Chem. Eng. News* **2013**, *91* (22), 39.
11. Wu, M. L. Vision 2025: Helping ACS Members Thrive in the Global Chemistry Enterprise. In *Vision 2025: How To Succeed in the Global Chemistry Enterprise Sponsoring Group*; Cheng, H. N.; Shah, S.; Wu, M. L., Eds.; ACS Symposium Series 1157; American Chemical Society: Washington, DC, 2014; Chapter 22.
12. Globalisation Trends and Their Impact on the Chemical Industry, Frost and Sullivan, 2011. www.slideshare.net/FrostandSullivan/globalisation-trends-and-their-impact-on-the-chemical-industry.
13. Spitz, P. H. *The Chemical Industry at the Millenium*; Chemical Heritage Foundation: Philadelphia, 2003.
14. The Outlook for the U.S. Chemical Industry, KPMG, 2010. www.kpmg.com/US/en/IssuesAndInsights/ArticlesPublications/Documents/us-chemical-industry-outlook.pdf.
15. Morawetz, M. Future of Chemicals III. The Commoditization of Specialty Chemicals, Managing the Inevitable, Booz and Company, 2010. www.booz.com/media/uploads/Future_of_Chemicals_III.pdf.
16. Reinhardt, C., Ed.; *Chemical Sciences in the 20th Century: Bridging Boundaries*; Wiley: New York, 2001.
17. Rogers, M. E., et al., Eds.; *Preparing Chemists and Chemical Engineers for a Globally Oriented Workforce: A Workshop Report to the Chemical Sciences Roundtable*; National Research Council (U.S.) Chemical Sciences Roundtable: Washington, DC, 2004.
18. Databases, Tables & Calculators by Subject, U.S. Bureau of Labor Statistics. http://data.bls.gov/timeseries/CES3232500001?data_tool=XGtable.
19. Rovner, S. L. Salary & employment survey for chemists. *Chem. Eng. News* **2012**, *90* (39), 40–43.
20. Morrissey, S. R. Starting salaries. *Chem. Eng. News* **2012**, *90* (23), 36–38.

21. ACS Office of Research & Member Insights, 2012.
22. Weissmann, J. The Ph.D Bust: America's Awful Market for Young Scientists—in 7 Charts, *The Atlantic*, February 20, 2013. http://www.theatlantic.com/business/archive/2013/02/the-phd-bust-americas-awful-market-for-young-scientists-in-7-charts/273339/.
23. National Public Radio (NPR), March 10, 2013. http://www.npr.org/2013/03/10/173953052/are-there-too-many-phds-and-not-enough-jobs.
24. Maslen, G. The Changing PhD – Turning out Millions of Doctorates, *University World News*, April 3, 2013. www.universityworldnews.com/article.php?story=20130403121244660.
25. Feldman, D. Hidden Job Market Secrets, 2014. www.jobwhiz.com/networking.php.
26. Basalla, S.; Debelius, M. *So What Are You Going to Do with That?*; University of Chicago Press: Chicago, 2007.
27. Fischer, K. U.S. Will Be the Fastest Growing Foreign Student Destination. *The Chronicle of Higher Education* **2013**, October 8.
28. Song, J. U.S. Colleges Draw Record Foreigners. *Los Angeles Times* **2013**, November 11.
29. Altbach, P. G.; Reisberg, L.; Rumbley, L. E. Trends in Global Higher Education Tracking an Aacademic Revolution, Report Prepared for the UNESCO 2009 World Conference on Higher Education, 2009. http://www.uis.unesco.org/Library/Documents/trends-global-higher-education-2009-world-conference-en.pdf.
30. Zambon, K. AAAS Analysis Shows Uncertain Future for Federal R&D Spending, 2013. www.aaas.org/news/aaas-analysis-shows-uncertain-future-federal-rd-spending.
31. For more on ACS career resources, see www.acs.org/careers .
32. Wu, M. L. Looking for a job? Check out these tools for chemists. *Chem. Eng. News* **2013**, *91* (42), 35.
33. Wu, M. L. Time to partner and speak up for science. *Chem. Eng. News*, *91* (35), 54.
34. For more on ACS advocacy efforts, see www.acs.org/policy.
35. For more on ACS international activities, see www.acs.org/international.
36. Cheng, H. N.; Jones, W. E. The Global presence of the American Chemical Society: Resources & opportunities. *Chem. Eng. News* **2013**, *91* (33), 32.
37. Cheng, H. N.; Morello, M. Helping members thrive through global connections. *Chem. Eng. News* **2013**, *91* (43), 37.
38. Wu, M. L. Partners for progress & prosperity. *Chem. Eng. News* **2013**, *91* (1), 2–4.
39. Wu, M. L. Partners for progress & prosperity. *Chem. Eng. News* **2013**, *91* (10), 33.

Chapter 2

The Power of Inclusive Innovation

Critical Success Factors for Science-Based Solutions

Douglas Muzyka*

**Senior Vice President and Chief Science and Technology Officer,
E. I. du Pont de Nemours and Company, Experimental Station,
Route 141 and Henry Clay, Wilmington, Delaware 19880, United States
*E-mail: doug.muzyka@dupont.com**

With the world's population projected to reach nine billion by 2050, the need to provide the food, energy and protection needs of people everywhere presents an unprecedented challenge. Historically, science-based innovation has made significant contributions to addressing this challenge. As we continue to confront the needs of the global marketplace in the future, solutions will require increasingly innovative scientific advancements. These advancements must be focused on new applications and integration of scientific disciplines executed through collaborative partnerships with industry, governments, and academic institutions. This article reviews what we believe at DuPont to be the three critical factors for successful and sustainable science-based solutions: (1) market-driven innovation; (2) integrated science; and (3) Inclusive Innovation. This article also describes examples of sustainable solutions from DuPont science and technology.

Introduction

The global population surpassed seven billion in 2011, and it is expected to grow by two billion people to reach nine billion by 2050 (*1*). That translates roughly into 150,000 new people added to the world every day.

This relentless population growth and shifting economic patterns (*2*) will result in significant increases in demand for food, energy and protection. For example, the Food and Agriculture Organization of the United Nations estimates that by 2050, demand for food will grow by 70 percent (*1*). Additionally, the U.S. Department of Energy projects that by 2035, demand for energy will rise by as much as 53 percent (*3*).

To meet these increasing demands, we must continue to increase agricultural productivity, identify new and diverse energy sources, and find new and better ways of protecting people and the environment. This will require an even more detailed understanding of our environment, creation and application of new elements of science, and new and creative ways of working collaboratively together.

DuPont has been at the forefront of leading-edge science and technology over the last 200 years, and we are applying our expertise in conjunction with others to make impactful contributions to society in meeting these demands.

Our Approach

DuPont is a science company with core capabilities spanning across a broad range of chemistry, engineering, material science, and biology. By combining our scientific expertise with our market reach into diverse industries in greater than 90 countries, we are uniquely positioned to be able to gain critical market insights and create scientific solutions that will have both global and local impact.

Our approach to creating solutions for these global challenges utilizes three critical success factors:

- *Market-Driven Innovation.* We believe that true innovation must deliver real value to our customers and meet their needs. We work closely with customers at all stages (*4*) to identify unmet needs, develop innovative solutions and bring those solutions to the market. As a global company, DuPont leverages its multinational capabilities and applies them to create local solutions.
- *Integrated Science.* Science is more powerful when it integrates knowledge across the traditional disciplines such as biology, chemistry, material science and engineering. This integration allows for new, more impactful solutions, and solutions that are more sustainable.
- *Inclusive Innovation.* Meeting the world's growing food, energy, and protection needs will require more solutions than any one individual or one company can provide. That's why we use the problem-solving power of Inclusive Innovation. Today, we're working with more people in more places than ever before to create new ways to address the world's future needs.

Increasing Food Security

Historically, science has had a significant impact on increasing agricultural productivity, with many of the most significant advances coming over the last 80 years. These advances are best illustrated by the increases in corn productivity per acre that have been achieved through science-based solutions.

From 1900 through the mid-1930s, agricultural productivity was around 30 bushels per acre each year (*5*). Advances such as the invention of hybrid corn in the 1920s (*6*), followed by new industrial fertilizers and tractors and other mechanized tools, together increased yields dramatically. The introduction of genetically modified corn provided even greater productivity increases, and together, these advances in science and technology have resulted in an increase in the average bushels per acre for North American corn today to around 160 bushels per acre (*7*).

Today, and into the future, an important part of agricultural innovation is the ability to provide products that meet the specific needs of the geography where the product is used (*8*), even in the most challenging growing conditions.

Drought Tolerant Corn

For example, drought can have devasting impacts on crop yield and is generally unpredictable. Using science to improve drought tolerance in crops is an important priority for DuPont.

While science hasn't been able to produce cactus-like durability in corn, we at DuPont are able to improve yields under water stress conditions. By utilizing advanced genetics and breeding capabilities, we are overcoming limited water conditions and growing corn more effectively with less water. Indeed, DuPont Pioneer has developed Optimum® AQUAmax™ corn hybrids (*9, 10*) which use less water per acre, and in on-farm comparisons against competitive hybrids showed an 8.9 percent yield advantage over the competitive brand under water-limited conditions. We believe that we can continue to improve drought tolerance through science, and we are applying our strength in biotechnology to develop even better drought tolerance protection (*11*).

Insect Control in Agriculture

Besides overcoming drought, controlling diseases, insects and other pests is vital to robust food production. For many years, farmers have relied on various crop and seed protection products. Today, and into the future, these products must be increasingly effective sustainable solutions that improve productivity, profitability and crop safety.

To protect crops against damage from insect pests which can cause significant losses, DuPont Crop Protection developed Rynaxypyr® for targeted insect control. Rynaxypyr® was the first of a new class of anthranilic diamide insecticides and has one of the most favorable safety profiles of all commercial insecticides, including insecticides derived from natural sources. Yet it is one of the most selective insecticides ever developed, which makes it an ideal replacement for many other products in global markets.

Rynaxypyr® is currently registered in 92 countries on 400 crops and products containing it are the fastest growing family of insecticides. In 2012 alone, more than 28 million farmers around the world benefitted from the insect control provided by Rynaxypyr® (*12–14*).

Biofortified Sorghum

The African Biofortified Sorghum (ABS) project (*15*) is another example of modern science enhancing food production. Sorghum, which has long been a staple of the African diet, is able to grow in the hot and dry climates that characterize much of sub-saharan Africa. Yet it lacks the minimal recommended daily allowance of certain vitamins and minerals that contribute to proper development and health in humans, especially in young children. In addition, native sorghum proteins are difficult to digest.

ABS is a project that seeks to provide a long-term solution for Africa by creating transgenic sorghum that grows well in the harsh climate of sub-saharan Africa and contains increased levels of stable pro-vitamin A, greater bio-availability of iron and zinc, and increased protein digestibility after cooking. DuPont established collaborations with the Bill and Melinda Gates Foundation and a number of Africa organizations, such as the Africa Harvest Biotech Foundation International, the University of Pretoria, the Council for Scientific and Industrial Research of South Africa, the Kenya Agriculture Research Institute, the Institute of Agricultural Research of Nigeria, and several others to advance this technology. This public-private endeavor recognizes that it will take collaborative efforts achieved through Inclusive Innovation and the expertise of many entities applying integrated science to tackle this nutrition and health challenge.

The end result of this work will improve the life and health of people dependent upon sorghum as their staple diet. Improved sorghum developed through the program will directly address the shortcomings in sorghum nutrition, having the potential to improve the situation for the estimated 250,000 to 500,000 children who become blind every year from vitamin A deficiencies in large parts of the developing world.

Conserving Energy and Developing Alternatives

While the demand for energy grows, the supply of fossil fuels will not increase. With a growing population, we will need to use existing resources as efficiently and effectively as possible, and we must find better ways to harness renewable energy sources as well. For instance, cars and airplanes need to be lighter and more fuel efficient, solar energy needs to become even more proficient at providing power at a lower cost, and the production costs of biofuels need to be aligned with the market. All of these goals are being enabled today by integrated science at DuPont.

Automotive Lightweighting

At the most basic level, improving vehicle efficiency reduces the amount of energy needed to power a vehicle. Therefore, a major thrust for DuPont in this area is to increase the fuel efficiency of cars through lightweighting by incorporating engineered plastics and composites into new vehicles. However, because today's higher efficiency vehicles rely on multiple performance-enhancing technologies to get greater horsepower from smaller engines, they create hotter, more chemically agressive environments than traditional plastics can withstand. DuPont polymer scientists and our market-facing development engineers collaborated with automotive companies to reinvent Zytel® nylon, which is now being used to replace metal under the hoods of vehicles (*16*).

DuPont is also actively engaged with many automakers, universities and consortia to help develop the next generation of lightweight composite materials. We understand that large scale adoption of lightweighting technologies that can significantly reduce vehicle mass will require a number of breakthroughs best achieved through collaborative efforts throughout the value chain.

Solar Energy Solutions

DuPont is already playing a leading role in the energy marketplace. For instance, we have been a very large supplier to the solar energy industry for many years, and our materials can be found in a majority of solar panels manufactured over the last thirty years.

Recently, a team at DuPont applied integrated science to further increase cell efficiency and reduce cost by focusing on improving the metal-silicon interface contact of solar cells. The result is a new chemistry platform that enables better electrical conduction. DuPont™ Solamet® PV17X was the first photovoltaic metallization paste based on this new chemistry platform. Advances in Solamet® metallization have doubled the efficiency of solar cells over the last 12 years, and are expected to continue to help the industry in its efforts to reach a new goal of 22 percent efficiency in 2015 (*17*).

Biofuel Solutions

Delivering low cost and scalable solutions that provide governments around the world a path to meet their renewable fuels goals, reduce greenhouse gas emissions, and create greater energy independence requires the integration of multiple scientific and engineering disciplines. To accomplish this, DuPont has taken a multi-pronged approach to making cellulosic biorefining a reality. It includes leveraging unparalleled knowledge of the feedstock supply chain through our DuPont Pioneer agronomy experts, a fully integrated technology package, and our knowledge and ability to produce key consumables such as enzymes and biological catalysts on an industrial scale and at the cost our customers demand.

The DuPont cellulosic ethanol process is a novel, integrated production platform with three major technology components for the production of ethanol at sufficiently high yields and titers to achieve commercially viable economics. To optimize the process it was necessary to integrate and innovate all three conversion steps.

Beginning with a close relationship with farmers, and through investing in novel collection systems, we control the precise collection of biomass while optimizing sustainable land use practices for the farmers. In the DuPont process, a novel dilute ammonia biomass pretreatment process decouples the carbohydrate polymers from the lignin matrix with minimal formation of compounds which inhibit subsequent fermentation, thus eliminating the need for costly "detoxification" steps which are common in other cellulosic ethanol technologies.

Next, an enzymatic hydrolysis step uses a novel suite of high performance enzymes specifically engineered by DuPont to depolymerize and hydrolyze both cellulose and hemicellulose to high titers of fermentable sugars in a single sugar stream.

Third, we redesigned and optimized the metabolic pathways of a recombinant bacterium, *Zymomonas mobilis*, to simultaneously metabolize both 6-carbon (glucose) and 5-carbon (xylose) sugars to efficiently produce ethanol at high yields and titers from the hydrolysate.

This unique integration of three distinct technologies enables a very efficient process with minimal steps, a reduced environmental footprint, and reduced cost of capital versus other known cellulosic ethanol processes. Today our engineering team is partnering with leading design and construction companies to build a first of its kind cellulosic ethanol plant in Nevada, Iowa, U.S.

Delivering Enhanced Protection for the Growing Population

According to the Centers for Disease Control, life expectancy for an individual in the U.S. has increased from an average of approximately 47 years in 1900 to about 78.5 years by 2009 (*18*). If we look around the world, we see a similar change. Over the years, scientific innovations have contributed to the upward climb in life expectancy. While pharmaceuticals continue to make a tremendous contribution, other products also support enhanced life expectancy, general human health, and a cleaner, safer environment.

Detection of Pathogens in Food

The DuPont BAX® System helps protect the quality and safety of our food. It is a fast and accurate system for detecting pathogens or other organisms in food and environmental samples (*19*). The BAX® System uses polymerase chain reaction (PCR) to amplify a specific fragment of bacterial DNA. For example, Salmonella can be detected reliability through the amplification of its unique genetic sequence, thus providing a highly reliable indicator when the pathogen is present. The BAX® System simplifies the quality assurance process by combining the requisite primers, polymerase and nucleotides into a stable, dry, manufactured tablet pre-packaged in tubes. After amplification, these tubes remain sealed for the detection phase, thus significantly reducing the potential for contamination with one or more molecules of amplified PCR product. The BAX® System measures the magnitude and characteristics of fluorescent signal change. An analysis by the BAX® System software algorithm then evaluates that data to determine a positive or negative result which is displayed and analyzed. This system helps preserve and protect the integrity of the world's food supply by detecting pathogens which can then be eliminated.

Auto Armor

DuPont's Kevlar® aramid fibers are used in a variety of clothing, accessories, and equipment to make products stronger and provide enhanced protection to those that rely on them. Kevlar® is lightweight and extraordinarily strong, with five times the strength of steel on an equal-weight basis (*20*). Since its invention over 40 years ago, Kevlar® is now used in everything from the backbone material for police and military vests to lightweight airplane material, fiber optics, city roads, tires and even cars.

In many parts of the world, violence remains a daily threat for average citizens. In Brazil, for example, the murder rate can reach 40,000 a year (*21*). Until recently, providing a family car with armoring, including bullet-resistant panels and protective window layers, was a largely an unrealized protection option.

In 2008, DuPont introduced an innovative, retrofit solution to make car armor more affordable in Brazil at more than 50 percent lower than market prices for other solutions (*22*). Called DuPont™ Armura®, this particular solution uses Kevlar® for all panels and SentryGlas® laminate layers with SpallShield® protection for windows and glass. The Kevlar® panels are light, flexible, and molded for a precise fit, making installation fast and reliable. The thin, clear glazing solution in SentryGlas® was first used for hurricane-proofing, as it retains glass shards at the moment of impact. The entire package adds less than 198 pounds to the total vehicle weight more than 50 percent less than traditional armor weight, which averages 450 pounds.

With decades of experience in the protection industry, DuPont remains innovative and motivated in the fight for personal safety and protection, and is bringing solutions which address real customer needs, such as Kevlar®.

Summary

We will continue to be challenged to meet the demands of our changing world. At DuPont, we believe that science, especially integrated science, will play an increasingly important role in our future. By working with more people in more places, and integrating knowledge across scientific disciplines, we will continue to find new and better ways to solve global problems. These solutions will be critical to enable us to provide for the food, energy, and protection needs of people everywhere.

References

1. How To Feed the World in 2050, Executive Summary. Food and Agriculture Organization (FAO) of the United Nations (UN). http://www.fao.org/fileadmin/templates/wsfs/docs/expert_paper/ How_to_Feed_the_World_in_2050.pdf (accessed July 30, 2013).
2. Preparing for the Future. U.S.D.A. Advisory Committee on Biotechnology and 21st Century Agriculture. http://www.usda.gov/documents/scenarios-4-5-05final.pdf (accessed July 30, 2013).

3. 2013 International Energy Outlook U.S. Department of Energy. http://www.eia.gov/forecasts/ieo/table1.cfm (accessed July 30, 2013).

4. DuPont Innovation Centers Worldwide. http://www2.dupont.com/Innovation_Centers/en_US/ (accessed July 30, 2013).

5. Are We Capable of Producing 300 bu/acre Corn Yields? Iowa State University Agronomy Extension. http://www.agronext.iastate.edu/corn/production/management/harvest/producing.html (accessed July 30, 2013).

6. Crow, J. F.; Dove, W. F., Eds.; Perspectives: Anecdotal, Historical and Critical Commentaries on Genetics. In *90 Years Ago: The Beginning of Hybrid Maize*, 1998. http://www.genetics.org/content/148/3/923.short.

7. Average U.S. Corn Yield Per Acre. Corn Farmers Coalition Charts/Data. U.S. Department of Agriculture, Economic Research Service. http://www.cornfarmerscoalition.org/fact-book/chartsdata/ (accessed July 30, 2013).

8. Löffler, C. M.; Wei, J.; Fast, T.; Gogerty, J.; Langton, S.; Bergman, M.; Merrill, B.; Cooper, M. Classification of maize environments using crop simulation and geographic information systems. *Crop Sci.* **2005**, *45*, 1708–1716.

9. Optimum® AQUAmax™ Products. www.pioneer.com/aquamax (accessed July 30, 2013).

10. Optimum® AQUAmax™ Hybrid Grower Systems Trials (2010-2012). https://www.pioneer.com/home/site/mobile/agronomy/aquamax-trials-western (accessed July 30, 2013).

11. Barker, T.; Campos, H.; Cooper, M.; Dolan, D.; Edmeades, G.; Habben, J.; Schussler, J.; Wright, D.; Zinselmeier, C. Improving drought tolerance in maize. *Plant Breed. Rev.* **2005**, *25*, 173–253.

12. Lahm, G. P.; Stevenson, T. M.; Selby, T. P.; Freudenberger, J. H.; Cordova, D.; Flexner, L.; Bellin, C. A.; Dubas, C. M.; Smith, B. K.; Hughes, K. A.; Hollingshaus, J. G.; Clark, C. E.; Benner, E. A. Rynaxypyr: A new insecticidal anthranilic diamide that acts as a potent and selective ryanodine receptor activator. *Bioorg. Med. Chem. Lett.* **2007**, *17*, 6274–6279.

13. Cordova, D.; Benner, E. A.; Sacher, M. D.; Rauh, J. J.; Sopa, J. S.; Lahm, G. P.; Selby, T. P.; Stevenson, T. M.; Flexner, L.; Caspar, T.; Ragghianti, J. J.; Gutteridge, S.; Rhoades, D. F.; Wu, L.; Smith, R. M.; Tao, Y. Elucidation of the mode of action of Rynaxypyr, a selective ryanodine receptor activator. *Pestic. Chem.* **2007**, 121–126.

14. Lahm, G. P.; Cordova, D.; Barry, J. D. New and selective ryanodine receptor activators for insect control. *Bioorg. Med. Chem. Lett.* **2009**, *17*, 4127–4133.

15. The Africa Biofortified Sorghum Project. http://biosorghum.org/home.php (accessed July, 2013).

16. Zytel® Plus High Temperature Plastic Resin. http://www.dupont.com/products-and-services/plastics-polymers-resins/thermoplastics/brands/zytel-nylon/products/zytel-plus-high-temperature-plastic.html (accessed July, 2013).

17. DuPont Photovoltaic Solutions.   http://www2.dupont.com/Photovoltaics/en_US/news_events/article20120920.html (accessed July 30, 2013).

18. 2011, Life Expectancy at Birth, at Age 65, and at Age 75, by Sex, Race, and Hispanic Origin: United States, Selected Years 1900−2010, Table 22. U.S. Centers for Disease Control. http://www.cdc.gov/nchs/data/hus/2011/022.pdf (accessed July 30, 2013).

19. Developing Faster, More Accurate Food Safety Tests for the Industry. DuPont. http://www.dupont.com/corporate-functions/our-approach/global-challenges/food/articles/food-industry-tests.html (accessed July 30, 2013).

20. Rebouillat, S. Aramids. In *High Performance Fibers*; Hearle, J. W. S., Ed.; Woodhead Publishing: Cambridge, U.K., 2000; pp 23−61.

21. United Nations Statistics Division.   http://data.un.org/ (accessed July 30, 2013).

22. DuPont   Armura®.   http://www.dupont.com/corporate-functions/our-approach/global-challenges/protection/articles/brazil-life-protection.html (accessed July 30, 2013).

Chapter 3

Chemistry at the Core of Biomedical Innovation

Alan D. Palkowitz*

**Vice President, Discovery Chemistry Research and Technologies,
Eli Lilly and Company, Lilly Corporate Center,
Indianapolis, Indiana 46285, United States
*E-mail: palkowitz_alan_d@lilly.com**

The success of the pharmaceutical enterprise during the last century in extending life expectancy and improving the quality of human health with novel therapies is unparalleled and has set the benchmark for future innovation. While scientific, business and regulatory challenges have limited the productivity of the industry during the past several years, key learnings are providing new paths to future discoveries that will reverse these trends and yield novel medicines that address unmet patient needs. Organizations that will contribute to this future will be distinguished by their ability to engage complex diseases with diverse strategies that are enabled by a talent base capable of solving difficult challenges at the interface of multiple scientific disciplines. One dominant theme is that the unique skill set of chemists will place them at the center of catalyzing biomedical discovery. However, this will require adaptation that challenges chemists to transcend mastery of their core disciplines in order to deliver on the growing demands of breakthrough innovation. Key to this future will also be unique models for collaboration that bring together diverse scientific expertise and capabilities from around the globe in a coordinated and focused effort. In this lecture, a perspective on the future of drug discovery and the opportunities for the scientific leadership of chemists will be shared.

In considering the broad question of how to succeed in the global chemistry enterprise of the future, it is important to include multiple frames of reference as chemistry touches many parts of science, business and society. While the present financial climate has clouded the prospects for employment in many chemistry-based industries, my strong belief is that the future will provide abundant opportunities, since the many disciplines of chemistry will remain the core transformative science for delivering tomorrow's innovation. In this chapter, a brief perspective of the role that chemistry has and will continue to play in biomedical innovation will be shared.

By any measure, the impact of the chemical sciences during the last century in shaping pharmaceutical research and development is striking. To appreciate this, it is instructive to go back even further, perhaps 4.5 billion years to the time associated with the early formation of the earth. Over that time span, a massive biosynthesis has occurred, creating the foundation of our natural world. Nature has produced an ever-growing, diverse collection of molecules that have been selected for evolutionary advantage and are uniquely associated with all forms of life. However, only in the last century have we been able to routinely study this molecular world through an understanding of chemical structure and function (1). Advances in theoretical and experimental chemistry and the associated analytical and spectroscopic sciences have provided us insights into the molecular architecture and mechanisms of biological processes. Likewise, synthetic chemistry has developed as a science that allows us to precisely create and modify chemical substances that alter human physiology and disease. It is no accident that progress in the field of chemistry has provided a gateway to contemporary drug discovery.

As an example, the concept of systematic modification of chemical structure through synthesis to elicit and optimize a desired pharmacological response was triggered by the groundbreaking work of Paul Ehrlich early in the 20th century. Ehrlich was an immunologist whose research on antibodies and the theory of immunology earned him the Nobel Prize in 1908. Through his work, Ehrlich made the observation that synthetic dyes produced differential staining of biological tissues. He reasoned that since some dyes could selectively stain microorganisms or specific tissues, they could be used to treat diseases caused by these microorganisms, or specific diseases of those tissues. In so doing, he invented the concept of chemotherapy and initiated several programs to explore this possibility (2).

Ehrlich's work on the discovery of Salvarsan, a treatment for syphilis, was an early example of the systematic approach to optimize the efficacy of a molecule by iterative structure-activity studies. In describing his research, Ehrlich stated that "we must learn to aim and to aim in a chemical sense" (3). While rudimentary by today's standards, Ehrlich successfully brought together several disciplines of science that are the foundation of modern drug discovery and opened new possibilities for the concurrently emerging field of synthetic organic chemistry. Additionally, Ehrlich was first to recognize the possibilities for conjugation of antibodies with cytotoxic agents as a vehicle for targeting active drugs with high specificity. Although not pursued at the time, this concept is an active strategy in contemporary drug discovery.

While the theoretical and applied examples of Ehrlich's work were groundbreaking, the societal impact of Salvarsan in transforming the treatment of syphilis was profound and representative of a prolific period for pharmaceuticals that would follow. Salvarsan demonstrated unprecedented efficacy in treating syphilis and became the most widely prescribed drug in the world until it was replaced by even more efficacious (and safer) antibiotics (4).

During the 20st century, countless additional discoveries in medicinal chemistry dramatically improved the quality of life for patients suffering from a vast array of diseases. These breakthrough therapies included insulin, penicillin and synthetic antibiotics, Taxol, Prozac, the statins, anti-HIV therapy, just to name a few. In many areas of disease biology research, the translational barrier for creating new medicines was lowered when chemists provided the breakthrough molecules necessary to test intervention hypotheses (5). Thus, it is well appreciated that the ability of chemists to expand this foundation will continue to be a source of pharmaceutical innovation.

With such a well-established precedent, one cannot escape the contrast between the great progress of the last century and the current challenges faced by the pharmaceutical industry. It is no secret that the pharma industry today is recovering from a tumultuous period of defining a new path forward. Many forces are exerting pressure on the industry, including rising costs and reduced productivity in R&D, difficulties with long development times and looming patent expirations, concerns over drug safety, and increasing pressure on pharmaceutical prices (6). These realities have left many asking the question, "With so much promise, what happened?" While the answer to this question is very complex, the consequences have been quite profound. The overall outcome of an ongoing series of mergers and workforce reductions has been the collective consolidation of the industry that has occurred over the past 20 years. This has dramatically reduced the number of independent centers of innovation that are capable of reversing the trends through revitalization of R&D. Unfortunately, for companies that have followed this path, near-term financial benefits have often given way to renewed pipeline challenges. Another tragedy, of course, is the impact on both the current scientific workforce and the potential future scientists that are establishing their career directions under these conditions. The confounding paradox is that while chemists have borne the brunt of these challenges through research staff reductions, they are, in fact, central to the future transformation of the pharmaceutical industry and its return to a more productive posture.

In looking toward the future and the roles for chemists in biomedical research, it is first important to highlight the opportunity that remains before us. The need for new medicines continues to be great. We face disturbing trends in the rise of diabetes, cancer, cardiovascular and neurodegenerative diseases. Our populations are aging and will live longer. Emerging global economies, such as those in Asia, will create unprecedented demand for pharmaceuticals. Additionally, we are in a period of continual scientific breakthroughs that are providing tremendous insight into disease and possible directions to improve on existing therapies as well as address unmet medical needs. These breakthroughs have also paved the way for consideration of new approaches to R&D that leverage technology and global collaboration in exciting ways. The key question is how we can make the most of these opportunities.

In order to create a different future, it is important to consider where we have been in the past 20 years and capitalize on key learnings. As mentioned earlier, this is a complex task due to the multiple environmental and non-technical factors that have shaped the industry over this period. However, one area that certainly influenced a global shift in the approach to drug discovery was the industry's response to the explosion of information associated with genomics.

The sequencing of the human genome greatly expanded the number of potential drug targets and shifted the focus of drug discovery to a reductionist mindset dominated by technology and process driven paradigms. Organizations scaled their size to pursue many discovery projects in parallel with the goal of rapidly identifying the next blockbuster agents. High throughput technologies industrialized aspects of both chemistry and biology. Processes, rules and filters were used to manage large volumes of data and facilitate decision making. Molecules were advanced into human study for targets whose connections to disease were poorly understood and in some cases were limited by safety and pharmacokinetic factors that did not allow for thorough clinical experimentation (7).

While there have been many notable successes during this period, the industry as a whole has fallen far short of delivering on its promise. I believe that many of the challenges we now face are resultant from a strategy that neglected an appreciation of how prior innovation was discovered and developed. In the past, chemical synthesis was often the rate-determining step for drug discovery. Great care was given to molecule design with in-depth biological characterization of each molecule, in order to maximize the value of this key resource. Paradoxically, the development of modern synthetic methods and high-volume biological testing approaches has caused us to "forget" the importance of these steps. The multi-factorial genetic complexity of human disease, coupled with the complex interactions of drugs in physiological systems is now leading to the recognition that a drug discovery approach driven by numbers alone cannot be successful. Going forward, an honest assessment of our understanding of diseases and patients coupled with reevaluation of our discovery approaches will be necessary to truly transform knowledge into value.

While many challenges remain, there are emerging positive signs that the strategic gaps of the past are being closed with newer paradigms that are beginning to populate clinical development pipelines with innovative molecules. As a leading indicator, the number of FDA new drug approvals in 2012 was the greatest in the past 10 years and represented 18 novel mechanisms across a broad diversity of disease indications (8). Incidentally, the vast majority of these medicines are small molecules designed and created by chemists. Also, there are roughly 8,000 molecules in clinical development that represent a healthy substrate for potential future therapies. Collectively, these encouraging data may represent a turning point for the industry and fundamentally bring back into view the importance of chemists in leading the return to delivering breakthrough pharmaceutical innovation.

If these trends remain consistent, a key question to examine is what has fundamentally changed? It is my belief that the answer lies in the scientific diversification of our approach to drug discovery. This is reflected in a realization that there are multiple paths to innovation that begin with a greater appreciation of the unique genetic backgrounds of the patients we wish to serve. Improvement in the successful translation of disease target hypotheses to desired clinical outcomes is encouraging approaches that tailor new medicines to patients who are likely to respond best based on unique markers of disease.

Furthermore, with a growing understanding of disease biology and refined therapeutic hypotheses, medicinal chemists are developing a range of diverse molecular discovery strategies along the continuum of small and large molecules that are expanding the drugability of the genome and increasing our success in creating novel agents for a growing diversity of drug targets. No longer limited by high-throughput screening as a dominant approach to drug discovery, this has been driven by the creative application of new biophysical and computational technologies that are dramatically increasing the sophistication of molecular design and optimization.

Additionally, scientific diversity and problem solving is being enriched by new collaboration models that are bringing together global researchers in creative ways to tackle some of the most difficult disease challenges.

In the remainder of the chapter, I will briefly expand on a few of these topics and share what I believe is a renewed path for chemists to transform biomedical science in the coming years.

In medicinal chemistry, multiple factors are dramatically shaping the field and creating new ways for chemists to evolve drug discovery to a higher level of practice. These include the challenge and opportunity of a growing diversity of therapeutic targets as well as a demand for both selective and multi-targeted agents. New insights into molecular design and mechanistic understanding of drug action are being driven by cellular and analytical technologies and increasingly sophisticated biophysical tools that allow us to "see" molecular interactions between synthetic ligands and proteins with unprecedented clarity. Computational and informatics tools are enhancing predictive modeling to expand the chemist's ability to evaluate multiple structural hypotheses and take on the challenge of parallel optimization of important drug properties. New strategies to fine tune the pharmacokinetic and pharmacodynamic responses of our medicines to optimize therapeutic index are being utilized to meet strict demands for product safety.

At this level, medicinal chemists are beginning to define solutions to problems that could open new areas of disease biology and bring forward an expanded range of therapeutic options for patients. These include the design of tissue-specific agents to improve safety and efficacy, merging of large and small molecules to create protein- and antibody- drug conjugates, and modulators of protein:protein interactions that have been largely intractable to previous small molecule strategies and could produce oral agents for diseases that have been limited to injectable biologics. For all of this new possibility, the basic skills of chemists are central to making it a reality. Just as Ehrlich revealed over 100 years ago, the ability of chemists to understand and master molecule structure, function, and properties through synthesis will be a key source of innovation for patients.

The importance of synthetic chemistry to this ongoing transformation cannot be overstated. The iterative cycle of drug discovery that links ideas to data is limited only by our ability to translate hypotheses into synthesized molecules. Thus, being able to put the right compound "in the bottle" is central to biomedical innovation today, just as it was in the 20th century. For any company, it is essential to have available the expertise and ability to make any compound we can imagine. While many technologies are helpful, none are more critical than knowledge of the contemporary art and practice of organic synthesis.

As backdrop to this point, some key trends in pharmaceutical synthetic chemistry over the past 20 years are worth highlighting. As described earlier, pharma took an unfortunate turn with the advent of high-throughput technologies in an attempt to industrialize the systematic screening of many genomic targets. Synthetic methods focused on increasing the number of compounds prepared in order to increase the odds for success. Physical properties and design methods were secondary to chemistries that allowed for the rapid construction of compounds imagined by combinatorial design arrays and limited to simple solid and solution phase synthetic techniques. In retrospect, for those in the industry who fully bought into this approach, what was produced during this period were large numbers of compounds with structural redundancy, generally poor physiochemical properties, and a high degree sp2 character and amide linkages. Very few successful drugs were advanced or delivered from this paradigm, a contributing factor to the slowdown of our industry during the past decade.

In contrast, drug discovery today is driven by insights and technologies that allow us to deconstruct the elements of biological activity and physical properties and then design the best possible ligands for clinical testing. An essential element in this approach, as noted earlier, is access to synthetic capabilities to make the molecules borne out of sophisticated experimental hypotheses. As drug molecules are becoming more complex, the availability of synthetic expertise of the highest order is essential. Without it, tomorrow's laboratory innovations will never make it to patients. Like many industries today, we must also be able to translate our discoveries with minimal environmental impact. This imperative in itself will place even more demands on chemistry innovation in the future.

In concluding this chapter it is important to return to a consideration of the individual chemist. While this piece has centered primarily on the role of organic chemists in pharmaceutical research, the major themes are broadly applicable to other chemistry disciplines as well. Many of the skills that chemists will need to succeed in the future are actually part of their fundamental training, although new skills that are developed through dynamic experiences in the changing scientific and business worlds are also critical. The ability to excel at hypothesis generation and experimental design as well as problem solving will always be core.

Additionally, and more than ever before, chemists in the future will need to work at multiple scientific interfaces and transcend their training to master associated scientific disciplines and technologies. Collaborative skills will be essential whether as a team member or working with global researchers as science becomes more of a networked enterprise. There are multiple forces acting on the public and private sector that are drawing researchers together on many levels to solve complex biomedical challenges. Central to this is a realization that no single group or individual can solve major innovation challenges alone and that collaboration models that unite global researchers are key to advancing science. This will certainly require chemists to be proactive in creating a visible presence in the scientific community. However, I believe that the basis for our collective success will ultimately be that chemists have the initiative and courage to engage and master scientific complexity. This is where innovations are created and the outcomes we all seek will follow.

Just as it was decades ago, the role of talented and motivated chemists in this mission will be undeniable. Further, we must ensure that our educational systems and academic research communities continue to produce future generations of scientists who will take on these challenges. My belief is that it begins with strong mentors and role models who share excitement about science and provide context for the role that chemists can play in improving human life and fundamentally changing the world. This is far less than certain in today's climate, but we all must find a way to address this together.

References

1. Wender, P. A.; Baryza, J. L.; Brenner, S. E.; Clarke, M. O.; Craske, M. L.; Horan, J. C.; Meyer, T. *Curr. Drug Discovery Technol.* **2004**, *1*, 1–11.
2. Kasten, F. H. *Biotech. Histochem.* **1996**, *71* (1), 2–37.
3. Erlich, P. *Ber. Dtsch. Chem. Ges.* **1909**, *42*, 17–47.
4. Thorburn, A. L *Br. J. Vener. Dis.* **1983**, *59* (6), 404–5.
5. Edwards, A. M.; Isserlin, R.; Bader, G. D.; Frye, S. V.; Willson, T. M.; Yu, F. H. *Nature* **2001**, *470*, 163–165.
6. Scannell, J. W.; Blanckley, A.; Boldon, H.; Warrington, B. *Nat. Rev. Drug Discovery* **2012**, *11*, 191–200.
7. Bunnage, M. E. *Nat. Chem. Biol.* **2011**, *7*, 335–339.
8. FY 2012 Innovative Drug Approvals. U.S. Food and Drug Administration.http://www.fda.gov/AboutFDA/ReportsManualsForms/Reports/ucm276385.htm.

Chapter 4

Maintaining a Strong Chemistry-Based Industry in the United States

Ronald Breslow*

Department of Chemistry, Columbia University, 3000 Broadway,
New York, New York 10027-2399, United States
*E-mail: rb33@columbia.edu

It is critical to the U.S. that we continue to be the home to advanced science and the technology-based industries that benefit from this science. In this talk I will describe three threats to our eminence in this area, particularly in chemistry and its applications. One is the need to attract more of our brightest students into the field, adding strength to America's future. Another is the non-economic effects from take-overs of pharmaceutical companies. The side effects are the weakening of our aggregate scientific strength--in many of the take-overs the strongest and most experienced chemists in the firm being purchased lose their positions and the opportunity to contribute their expertise to our science-based future. This latter problem also discourages our brightest students from entering a profession with such an uncertain career path ahead. A third related problem is the outsourcing of chemistry jobs in industry away from the U.S. Possible remedies to these problems will be discussed.

Whereas globalization is an important trend, an equally important action for us is to retain our strengths in science, technology, engineering and mathematics (STEM) within the United States. There are three problem areas that I would like to address in this chapter.

Greater Support for STEM Higher Education

First of all, the general public does not understand the critical role that science, particularly chemistry, plays in our society. Chemistry is a creative science. We invent new molecules, and we apply these molecules to applications. Probably two-thirds of chemistry graduates work in industry. Many of them do applied research that has immediate or longer term impact on our daily lives. In this way, chemistry is different from astronomy (for example), where synthesis is not possible. We appreciate nature, and we find out why grass is green or why a certain co-enzyme has the shape that it has. We ask how else we can learn from nature, and how we can mimic nature and synthesize new things. Thus, we are nature explorers and nature extenders, and we extend our science into technology and innovation.

In the U.S., chemistry research in academia is mostly performed by graduate students. A current problem is the difficulty of supporting them. At one time, it was easy for graduate student to get NSF and NIH fellowships for their own support. A few years back, I had 19 graduate students and 18 of them were supported by their own federal fellowships. However, times have changed. Now it is difficult to get enough funding to support the group sizes that were previously pretty large. I have seen young faculty members who have good ideas and work hard but who do not get tenure because of the difficulty of getting money to support their research programs. This is a real problem for academia.

In the U.S. education up to high school is supported by the Department of Education. We need to recognize that education does not stop at high school. The Department of Education should provide support all the way through graduate school, because education at these higher levels has greater impact on U.S. industrial competitiveness. There is currently an excellent program at the Department of Education called "Grants and Aids in Areas of National Need" (GAANN) (*1*). Certainly STEM can be considered areas of national need and should be funded under this program. The problems is that right now its budget is limited -- only a tiny fraction of the budget of the Department of Education. I suggest that we increase its funding to cover the STEM graduate students' stipends. Furthermore, the current GAANN program is run like a poverty program and graduate students have to be almost starving to qualify. This aspect of the program also needs to be rectified.

I understand that in China graduate education is supported by the government's education budget. This seems to be a wise policy. In the U.S., a big chunk of the research funds is used to pay for the graduate students' stipends. If the U.S. Department of Education can pay the graduate students' stipends, then the professors' research funds can support postdocs, purchase equipment, and cover research expenses, with better outcomes.

Incentives for U.S. Students To Study STEM

The second problem we have is that many of the best and the brightest students in the U.S. are not studying STEM. Instead, they prefer to enroll in law schools or medical schools. For the U.S. to be competitive in the world, and for U.S. chemical industry to be strong, we need to attract our best students to science in general, and chemistry in particular.

Something that has been detrimental to chemistry is the consolidation in the chemically related industries, especially pharma. Big companies are buying up smaller companies. Now I understand the logic of acquisition. I used to own a small company and raised $70MM in private equity to sustain it. However, I had other interests and did not want to run the company, so I sold it to Merck. It was a win-win situation. Merck could not invent all the things they wanted and needed to acquire new technology. I was freed from the operational part of the business and was glad that Merck took it over. Everyone was happy.

The problem happens when a big company buys a competitor. As we know, we live in a capitalist society and the essence of capitalism is competition. Acquisitions like that decrease competition and have a negative impact on innovation. They can also be detrimental to worker morale. I happen to know a good medium-sized company that was bought by a larger company. After the acquisition, the director of research lost his job because they did not need two directors of research. Even the second-level managers lost their jobs as well. Thus, they lost a lot of expertise and experience in this acquisition. The people fired had good management skills and product development knowledge, but they no longer have the opportunity to contribute their expertise to our science-based future. They can serve as consultants, but that is not the same thing. I was disturbed about this trend and even consulted with Glenn Hubbard, the head of Columbia University Business School. He explained to me that in general business acquisition is not bad *per se*: it increases corporate efficiency and expands corporate expertise. I concluded that consolidation is probably beneficial to the corporations, but detrimental to the nation's scientific development. Certainly, it is wasteful in terms of scientific talent, and also discourages bright students from studying chemistry.

How to reverse this trend? This will not be easy. We need political actions to decelerate consolidation. We need incentives to encourage more students to study chemistry. The availability of fellowships will help. If the Department of Education can provide funding for STEM graduate students, that will help.

Outsourcing

We know that outsourcing is taking place, and many people have talked about it. Outsourcing does not necessarily have to be bad; it can be a two-way street. For example, the U.S. is happy to take some of the business from Swiss companies. However, Switzerland has not outsourced everything; they have kept some of the businesses and jobs at home, and they have good programs to train their own students. We should learn from the Swiss.

Right now, many graduate U.S. chemistry programs have a lot of graduate students from Asia. This is fine, but many of these students want to go back home after they have been trained. Thus, if we depend on foreign students to do the research for us, we lose the talent when they return home. Ultimately we end up losing as regards to our relative scientific strength in the world.

What we need is a decent graduate policy to encourage bright U.S. students to study STEM, such as graduate fellowships and an expanded GAANN program as described above. As companies notice the excellent pool of talent within the U.S., they will be less likely to outsource research overseas. Switzerland is a good example. The Swiss companies may be global in their operations, but they maintain a strong presence at home. They continue to support education at home and keep their workers employed.

Thus, while globalization is real and we need to consider the impact of internationalization on our activities, we should not lose sight of our domestic needs, particularly our need to continue to be a leader in science and technical innovation.

Summary

What are the assets that the U.S. has? Yes, we have natural resources. However, it is our STEM expertise that has really made the difference historically. Even now, we know that we are still strong in molecular biology. It started in the U.S. and we still have a scientific and technological lead. The problem is that the public does not understand how important STEM is to our society. We need to get in touch with the Congress; the message is that chemistry is important to this country but the funding mode is far from ideal.

More specifically, we need to let the Congress know that more money should be allocated to the Department of Education in order to support STEM graduate students. I once discussed my ideas with the director of chemical sciences at NSF and he loved the ideas. I have written an editorial at C&EN on this topic as well (2). I think this would be a very useful and important initiative that would ultimately help to maintain a strong chemistry-based industry in the United States.

References

1. Graduate Assistance in Areas of National Need. U.S. Department of Education. http://www2.ed.gov/programs/gaann/index.html.
2. Breslow, R. *Chem. Eng. News* **2011**, *89* (24), 3.

Chapter 5

Chemists: Public Outreach Is an Essential Investment of Time, Not a Waste of It

Paul J. Bracher and Harry B. Gray*

Beckman Institute, California Institute of Technology, Mail Code 139-74,
1200 E. California Blvd., Pasadena, California 91125, United States
*E-mail: hbgray@caltech.edu

In this chapter, we discuss the state of the public image of
chemistry and some of the potential consequences of its
deterioration. We explain why it is important that chemists
engage the public to educate citizens about science and
communicate the value of scientific research. Next, as an
example of how chemists can interact with the public in a
meaningful manner related to their research, we discuss the
development and implementation of a multifaceted outreach
program by our group at Caltech. We close with suggestions for
how chemists can create their own outreach activities and how
our field can encourage this work by recognizing and rewarding
it.

The global chemistry enterprise, while historically strong, faces a growing
list of serious challenges. The pharmaceutical industry finds itself struggling to
maintain its former pace of drug discovery (1), federal funding for academic
research is becoming increasingly scarce (2), and the employment market for
chemists in the developed world may never have been tougher than it is now (3,
4). Despite these challenges, one cannot deny the prospect that it is chemists
who are positioned best to solve several of the most profound problems currently
facing humanity. Chemical catalysts will be required to efficiently convert the
energy in sunlight into safely-transportable fuels that meet our growing demand
for consumable energy (5, 6). It is chemists who synthesize the drugs—be they
small molecules or large proteins—that are still often the most effective and
least invasive means of treating many diseases. And the solution to our planet's

greatest historical mystery—how life originated on Earth roughly four billion years ago—is necessarily chemical; physics is too abstract to offer answers to the problem, and by the time one progresses into biology, the problem has already been solved.

Our ability to tackle these complex problems requires our continued commitment to outstanding research and the rigorous scientific training of the next generation of chemists. But given this emphasis, we fear that research chemists are becoming increasingly single-minded, further isolating themselves within the scientific community. We must not lose sight of the fact that we have a responsibility to communicate chemical knowledge and the importance of our research to the public, who generously supports it. If the field of chemistry is to maintain its health through 2025 and beyond, we must not overlook the importance of public outreach.

Troubling Times for the Public Image of Chemistry

The benefits that the chemical enterprise has bestowed on society are profound and innumerable. Molecules and materials developed by chemists are cornerstone drugs, fabrics, soaps, preservatives, disinfectants, dyes, and fuels. Our field has produced technology that is pivotal to industries like construction, forensic investigation, petrochemicals, electronics, national defense, agriculture, and medicine. The American Chemistry Council estimates that more than 96% of all manufactured goods are directly impacted by the business of chemistry, and the chemical enterprise supports nearly 25% of the United States' GDP (7).

With all of these grand achievements, one would think that chemistry, chemicals, and chemists would be held in the highest regard possible by society. However, despite the manifest benefit of our field to civilization, chemistry is currently plagued by serious problems with its public image (8–10). Our brand has steadily deteriorated from a zenith of "better living through chemistry" in the 1970s to the ever-worsening current climate where the word "chemical" brings with it the baggage of a despicable connotation (11, 12). It is now common for the public to view any chemical as bad—unnatural, toxic, and dangerous (13). Somehow, one of our most simple terms—and one that lies at the very foundation of our profession—has been perverted into something dreadfully sinister.

In today's retail stores, it is common to find products—including various cosmetics, fertilizers, and sunscreens—that are nonsensically and falsely advertised as "chemical-free". Marketing agencies know that they can play to the public's growing distrust of chemicals, and it works. While there are certainly cases where specific chemists and chemicals have had regrettably negative consequences for the public (e.g., thalidomide and the Bhopal disaster), surely these cases are offset by the numerous ways that chemistry has improved modern life. From the countless pharmaceuticals that fight disease to the wide array of synthetic materials found in technological gadgets, chemistry has made modern life possible. But while we know that chemistry, chemicals, and chemists are responsible for these advances, the average person-on-the-street does not associate them with our science. It would seem that, to the public, chemists don't make drugs—the pharmaceutical industry does. Chemists don't stop food from spoiling—the agricultural industry does. Chemists don't discover new materials for the latest fashions—the textile industry does. When it comes to technology, chemistry seems to get none of the credit and all of the blame. We need to do a better job of educating the public so we can sing the praises of our field and the ways in which chemistry makes possible all of the modern products and industries valued by society (*10, 14–16*).

The Importance of Chemists' Participation in Public Outreach

Research chemists seem to have a serious lack of motivation with regard to engaging the public. On one end, we do very little to communicate the benefits of our work. On the other end, we do little to push back when Madison Avenue and the media drag some of our most identifiable terms through the mud (*11, 17*). The chemical enterprise has stood by and done nothing to mount a serious challenge to the concept of "chemical free". Such apathy can fuel a vicious cycle, because if we decide to take the high road and not fight back, the negative slogans—whether accurate or not—will stick and further damage the public image of chemistry.

Organizations like the International Union of Pure and Applied Chemistry (IUPAC) and the American Chemical Society (ACS) have supported various initiatives and programs aimed at public education and outreach, but none of them has been able to take hold with a large population of research chemists. It would seem that research chemists are content to leave most of the work to specialists and enthusiasts in education and outreach, with former ACS President Bassam Z. Shakhashiri being a shining example, and perhaps, the most popular chemist among these champions. But the effort required to make a significant impact with the public seems too great for the shoulders of just a few individuals. Rather, it is time that *all* chemists—including research chemists— stepped forward to do their part.

One can argue that worrying about public education is a waste of our time as bench chemists—an unnecessary distraction from solving the world's problems through research. What is it to us if most of the population either knows nothing or has the wrong impression of what we do? After all, *we* know the truth and our chemistry will work regardless of whether the public thinks it is important.

The problem with that argument is that we live in a democracy. For a democracy—where the People govern by voting—to function efficiently, the electorate must be educated and informed. The steady decline of chemistry's public image is a massive problem, because it erodes support for our field. Taxpayers fund our research, and if they are convinced that not only is chemistry not helping the world, it is hurting it, then what is to stop politicians from eliminating funding for chemical research? Unfortunately, these cuts are already on the agenda of Congress. The leadership of the House Committee on Science, Space, and Technology recently moved to side-step the peer review process by blocking support for certain areas and types of research (*2*).

One could also argue that chemistry's problems of public distrust and nonsense like "chemical-free" advertising are not *our* problems (as research chemists); they are failures of our system of primary education. That may be true, but if chemistry loses hold of its brand in the sphere of public perception—regardless of how or why this happens—chemists will ultimately have to live with the consequences. The problem is only getting worse as chemists stand on the sidelines (*12*). We know what happens when chemists largely remain disengaged; it is time to run the experiment where chemists participate in public outreach on a larger scale.

The Power of Research To Engage Youth

We have discussed the importance of chemical research to solving some of the greatest problems faced by humanity, as well as the necessity that chemists communicate the importance of research to the public rather than hermetically sealing themselves within their laboratories. It is interesting to note that in our experience, not only is chemical research important for solving world problems, it is a wonderful vehicle for inspiring young people.

It might seem as though college freshman and younger students are too inexperienced to understand or appreciate university research, but we have found that high-powered science is particularly appealing to these students. An introductory class on chemical research, in which professors describe their work and give tours of their labs, has been a wild success at Caltech and has attracted many students to enroll as chemistry majors. Our research group at Caltech believes strongly in the importance of undergraduate research, and we have always hosted a large crop of Summer Undergraduate Research Fellows (SURFs). Beyond undergraduates, we have found that high-school students can also be very motivated to learn about—and participate in—advanced chemical research. While early college and high-school students might seem too inexperienced to be useful in a research laboratory, they are usually full of motivation and energy. Their presence increases the vibrancy of our lab, and their questions force their mentors to know their own research inside and out. We have had fantastic experiences working with young people of all ages, and we describe some of these programs below.

Outreach Efforts at the Center for Chemical Innovation in Solar Fuels

The Center for Chemical Innovation in Solar Fuels (CCI Solar) has taken an approach that integrates public outreach with a robust program of chemical research. We are an NSF-funded consortium of 13 universities and 19 laboratories dedicated to the discovery of materials to enable the efficient conversion of the energy in sunlight into fuels that can be safely stored and transported (*18*). Solar energy appears to be the best option to meet the future's global demand for consumable energy in a method that will not concomitantly pollute the environment (*19*). Our focus has been the use of sunlight to split water into molecular oxygen and hydrogen, and our teams work towards discovering materials for the construction of a photoelectrochemical cell that effects the (i) efficient evolution of hydrogen at its cathode, (ii) efficient evolution of oxygen at its anode, and (iii) separation of the products via an ion-permeable membrane.

CCI Solar prides itself not only on conducting outstanding scientific research, but on educating the public about the importance of our mission and the principles behind it. We have developed a robust program of public outreach that focuses predominantly on engaging youth from ages 8 to 18. The kids share our vision for the importance of discovering solar fuels to power the planet, and we have channeled their enthusiasm and enlisted their help in our "Solar Army" (see Figure 1). We describe these efforts below.

Figure 1. The official logo for our Solar Army of student researchers and their mentors. (Design by Christian Welch, a SHArK student at Caltech in the summer of 2010.)

Solar Energy Activity Lab (SEAL)

In 2009, Bruce Parkinson and coworkers described a system that could be used by amateur scientists to screen metal–oxide compounds for activity in the photoelectrolysis of water (*20*). The kit modified an electrochemical method they previously reported (*21, 22*) such that the method could operate on a Lego Mindstorms® platform—available at most toy stores—with supplementary parts supplied by Parkinson's Lab. The operation of the kit is simple. A glass plate with a conductive layer of fluorine-doped tin oxide (FTO) is patterned with mixtures of metal salts, typically nitrates. The plate is then heated in a furnace or kiln to generate patterns of metal oxides that can be screened for activity in the light-driven reduction of protons or oxidation of hydroxide. The students clip their plate to a lead connected to a rudimentary potentiostat, then immerse the plate—now a working electrode—in a solution of electrolyte. A small, constant bias potential is applied to the system, and spikes in current are monitored as a green laser pointer is scanned across the plate by a robotic platform built with Legos. The software reports a false-color image that maps the recorded current against the position irradiated on the plate. The "hottest" spots on the false-color image indicate prospective catalysts for further investigation.

This system became the basis of Parkinson's Solar Hydrogen Activity Research Kit (SHArK) and the SHArK Project (*23*). The SHArK Project, initially funded by The Camille and Henry Dreyfus Foundation, matches teams of (roughly) six students with a graduate student or postdoc mentor to search for catalysts for splitting water. Together, the team builds the instrument using the Lego Mindstorms® kit, then uses it to investigate metal-oxide catalyst space. Results can be shared online and used to flag catalysts that show promise for more detailed investigation by university researchers.

When Parkinson joined CCI Solar and moved to the University of Wyoming, our group at Caltech instituted the SHArK Project in a number of local high schools in Pasadena. Soon after, building on the pioneering work of Parkinson, Jay Winkler (Caltech) designed and constructed a second-generation system that simplified the experiment (*24*). Using an array of LEDs to irradiate spots of catalysts rather than a laser that sweeps across the plate, the new system decreased the time required for scanning from several hours to several minutes. This Solar Materials Discovery kit quickly became our Solar Army's weapon of choice, and the system was renamed the Solar Energy Activity Lab (SEAL).

Figure 2. A team of students in the SEAL program presents results in a symposium on the campus of Caltech. (Photograph by Carolyn Patterson.)

In the SEAL program, graduate students and postdocs at Caltech volunteer to serve as mentors for SEAL teams at high schools in the Los Angeles area. Schools are paired with one or two mentors, who guide teams of roughly six students in screening metal oxides as potential catalysts. Meetings typically take place once a week, and at the end of the school year, all of the students are invited to a conference at Caltech (SEAL-CON) to present their results (Figure 2). A major aspect of the success of the program is that the students get close-quarters interaction with an experienced researcher who has expertise in electrochemistry. The response to the SHArK and SEAL projects has been fantastic. Whenever H.B.G. gives talks around the country about the work of CCI Solar, e-mails quickly pour in from college professors, high-school teachers, and parents wondering how they can get their hands on a kit and become involved in the Solar Army (Figure 3). In some regards, the program is a victim of its own success. We find ourselves having to turn down excited prospective participants because we can't recruit mentors and manufacture the kits fast enough.

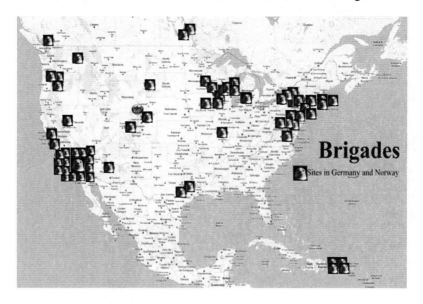

Figure 3. A map of the student brigades in our Solar Army involved in the SEAL and SHArK programs. (Graphic by Bruce Brunschwig.)

The SHArK and SEAL kits enable a form of crowdsourced citizen science, in that a large, labor-intensive problem can be broken down and distributed among many researchers (*25*). While there are a limited number of pure metal-oxide compounds to screen, there are an infinite number of higher-order combinations, e.g., 1:1 (Fe:Ni)O_x or 1:3:2:1:2 (Fe:Co:Er:Ca:Bi)O_x (*20*). By operating the user-friendly kits, high school students can perform meaningful research on the exact same problem as the principal investigators in CCI Solar. The kids who participate in SEAL don't see themselves as simply doing a science project; they see themselves as helping to save the planet, and they are absolutely right. While their chances of finding a catalyst to solve the world's energy problems are relatively small, the students are legitimately in the running. SEAL is not a laboratory exercise with a known outcome; it is real research.

Juice from Juice (JfJ)

Juice from Juice is a collection of educational activities for teaching students about dye-sensitized solar cells (DSSCs) and related topics under the umbrella of solar energy (*26*). Whereas SEAL is a platform for research projects meant to be conducted on a continuing basis by a small team of students, Juice from Juice is a set of lab exercises meant to be performed by full classrooms of students led by a high-school teacher. Using a kit designed by our team at Caltech, students can construct their own DSSCs using crushed up berries as a photosensitizer for titanium dioxide.

We have designed modules that are suitable laboratory exercises for classes in biology (e.g., photosynthesis), chemistry (e.g., electrochemistry), and physics (e.g., circuits and electricity). Sets of lesson plans in each subject—targeted for both middle-school and high-school audiences—are available on the program's Web site (*26*). We have also partnered with Arbor Scientific, a laboratory supply vendor, to make the kits available commercially.

In terms of implementation of the program, our team at Caltech runs workshops for interested teachers to try out the modules and learn how to teach them effectively. The teachers can then purchase the materials and use them independently in their own classrooms. Mentors from Caltech sometimes also run sessions themselves at science museums or as parts of science festivals in the Los Angeles area and beyond.

The Juice from Juice program is designed to engender wide engagement of children from relatively limited involvement of scientific mentors. In this model, a team of Caltech mentors can spend half of a day running a workshop for 20 teachers, each of whom can then teach the modules to five classes of 30 students. This model contrasts with the SEAL program described above, where mentors leave campus and visit schools one afternoon a week to work with the same small team of students for a full school year.

Informal Science Education: A Science Club for Kids

In 2012, we were approached by the National Science Foundation to extend our outreach efforts into the area of informal science education (ISE). ISE includes educational scientific activities that fall principally outside the boundaries of school curricula. Common examples include exhibits at museums, activities at camps, shows, clubs, science cafés, toys, and do-it-yourself experiment kits.

We partnered with the Westside Science Club (WSSC), an organization founded in 2008 by Benjamin Dickow, and the Wildwood School, a private high school in western Los Angeles. WSSC runs hands-on, inquiry-based science activities for children of ages 8–14 years old in the Venice neighborhood of Los Angeles. Our collaborative project is unique in that it leverages the community connections of the Westside Science Club—and its partner, the non-profit Venice Community Housing Corporation (VCHC), a low-income housing provider—with the STEM knowledge of chemists at Caltech and the enthusiasm of a team of local high school students that serve as "near-peer" mentors for club participants. The program meets every other Saturday morning in a community room located in one of the VCHC complexes and draws a regular group of up to 17 children.

The Westside Science Club places a premium on activities that promote skills, creative thinking, and problem solving above the simple acquisition of knowledge. The partnership with CCI Solar has maintained this philosophy, with Caltech researchers and WSSC facilitators working together to develop engaging experiences that go beyond simply talking to kids about science. The participants actively explore the foundational STEM concepts of CCI Solar's work through hands-on, inquiry-based activities and field trips, including a visit to the research labs on the campus of Caltech. The scientific activities in the first year of our collaboration have focused primarily on chemistry, with special emphasis on concepts associated with renewable energy—the expertise of the CCI research team. Figure 4 shows a "mind map" of concepts that graphically organizes single ideas (e.g., wavelength) into larger categories (e.g., light) and shows how the concepts are related. When planning meetings, we pick activities that highlight concepts that can be linked to ideas covered at previous meetings. For example, we linked the combustion of fossil fuels and release of carbon dioxide (club meeting #13) to global warming and the acidification of oceans (meeting #11). Each meeting begins with a review of the material from the previous session.

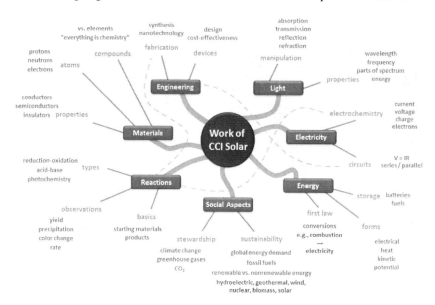

Figure 4. A "mind map" of some of the concepts that lie at the heart of understanding CCI Solar's research towards developing a photoelectrochemical cell for splitting water into molecular hydrogen and oxygen. The activities in our partnership with the Westside Science Club are designed to target these concepts.

45

Specific examples of activities from our first year included (i) using lantern batteries to produce hydrogen gas—a clean burning fuel—from the electrolysis of water, (ii) making glow sticks to demonstrate the concept of converting potential energy in reactive chemicals into light, (iii) using soda bottles and dry ice to discover how carbon dioxide can acidify oceans and cause an atmosphere to retain heat, and (iv) exploring the chemical properties of table salt and sugar (see Figure 5). The hallmark of these activities has been the hands-on nature of the experiments. The Caltech, WSSC, and near-peer mentors have worked together to design engaging experiments that the students can run safely by themselves, rather than sitting back and only watching demonstrations run by adults. Our approach favors active discovery of concepts rather than passive observation. In performing experiments like constructing homemade glow sticks from commercial chemicals, the students develop important laboratory skills, like transferring solvents by pipette and recording observations in a scientific notebook. The students also develop good habits with regard to safety, gain familiarity performing chemistry experiments such that chemicals are not "scary," and build confidence that science is a vocation that is both accessible to them and fun. Over the first academic year of the WSSC/CCI Solar collaboration, we have developed detailed lessons plans for the club's activities and documented the meetings such that the program can be replicated at future locations without having to start from square one. Indeed, plans for a second site in the Los Angeles area are already in the works.

Figure 5. Members of the Westside Science Club make colored solutions of table salt and sugar for a set of experiments that explores the properties of these compounds. (Photograph by Carolyn Patterson.)

How Chemists Can Get Involved in Outreach

Our vision is one in which future scientists—all scientists—spend a portion of their time in meaningful interaction with the public. We discuss some ideas and sources of motivation, below.

The Possibilities Are Endless

Effective outreach is by no means limited to CCI Solar's activities described above, nor will it necessarily resemble them. There is no limit to the activities and modifications available to those who want to step up and engage the public, and any activity where chemists bring science and chemistry to a population not already intimately involved in the field will make a significant contribution. Other potential activities could include: (i) judging a science fair at an elementary school, (ii) writing letters to the editors of newspapers to correct misinformation that has appeared in print (*14*), (iii) organizing or volunteering at a local science club, (iv) explaining your research at a science café, (v) volunteering to talk to a middle-school science class, (vi) developing a lab exercise for high schoolers based on your research, (vii) making a science Web site for a general audience, (viii) making YouTube videos (*9, 27*), and (ix) editing chemistry articles on Wikipedia (*28*). If you don't think you have a good idea, there are plenty of existing ideas in need of volunteers. The possibilities are endless!

The Rewards of Public Outreach: Existing and Suggested

Aside from the benefits to the individual citizens who participate in outreach activities, there are many benefits of outreach to the scientists who perform it: (i) With all of the extra practice interacting with laymen, scientists will become more adept at communicating and teaching technical material to nontechnical audiences (*29*). (ii) Chemists can also feel good about teaching people something new; outreach is a way to make the world smarter, one person at a time. It is also worth noting that education has always been something universally valued by our field. (iii) Scientists who participate in outreach with youth can feel good about giving back to society—many scientists owe their pursuit of the field, in part, to a role model or teacher who introduced them to science at an early age. (iv) In thinking about fundamental concepts and how to explain them, scientists will invariably come up with new ideas. Sometimes it's nice to think about areas of chemistry outside of your focus of research. (v) Scientists who participate in outreach will help to improve the funding climate by persuading voters that chemistry is a worthwhile activity.

If our field values public interaction, we must find ways to actively encourage it. Outreach has never caught hold in chemistry in the same way it has in the aerospace industry or medicine. We fear that one potential hurdle facing the growth of outreach in the field of chemistry is that many chemists view the activity as a waste of time with no tangible rewards. In terms of active encouragement, funding agencies like the NSF are already making leaps forward in advancing outreach by emphasizing the seriousness of the "broader impacts" evaluation of grant applications (*30*). We think that university programs should consider requiring some form of outreach involvement as a condition for the completion of any degree in chemistry. Perhaps, a chapter on outreach could also be a small component to any thesis or dissertation in the field.

In terms of passive encouragement, the culture of our field must change to embrace outreach efforts. Because outreach has been largely ignored by chemists for so long, the culture of our field does not particularly value it. If this is to change, the leaders of our field must lead by example and embrace outreach as a worthwhile activity. The outreach aspirations of junior faculty must be encouraged by those tenured in their departments, and research advisors must encourage and support their students and postdocs in this regard.

Conclusions

Chemists have a responsibility to engage the general public in discussions of science and scientific research. A healthy democracy requires an informed electorate, and in many cases, scientists are in the best position to educate the public about matters of policy. Our strong belief in the importance of public interaction is reflected in the considerable effort we have devoted towards a variety of outreach projects related to our research in solar energy, and we encourage our colleagues to get involved in any way they feel motivated. Public outreach is an essential investment of time, not a waste of it. The future of our field, in 2025 and beyond, depends on having an educated populace that appreciates the value of chemistry and chemical research.

Acknowledgments

Our research on solar fuels and the outreach efforts that accompany it are supported by the NSF CCI Program (CCI Solar, CHE–1305124). We thank the many scientists in CCI Solar who power our outreach efforts with their dedicated service as mentors in local schools and clubs. CCI Solar's outreach efforts are coordinated and managed by Carolyn Patterson and Siddharth Dasgupta. The SHArK and SEAL programs owe their success to the instrument platforms developed by Bruce Parkinson, Jay Winkler, and their coworkers. The SEAL program has been coordinated at Caltech by Jillian Dempsey, James McKone, Hill Harman, and James Blakemore, and on an international scale by Jennifer Schuttlefield. The Juice-from-Juice kit was developed by Qixi Mi and Michael Walter (Caltech), along with Debbie Hawks (Blair High School) and Gurupreet Khalsa (Pasadena Unified School District). The project has subsequently been coordinated at Caltech by Shane Ardo, Tania Darnton, and Amanda Shing. CCI Solar's ISE efforts have been spearheaded by Benjamin Dickow (WSSC), Ariel Levi Simons (Wildwood School), Anna Beck (Caltech), and P.J.B. (Caltech). P.J.B. gratefully acknowledges an NSF American Competitiveness in Chemistry postdoctoral fellowship grant (CHE–0936996). We also thank ACS President Marinda Li Wu, H. N. Cheng, and Sadiq Shah for organizing the Vision 2025 symposium and for the opportunity to discuss these important issues that face the chemical enterprise.

References

1. *Research and Development in the Pharmaceutical Industry*, Congressional Budget Office, Congress of the United States, October 2006.
2. Widener, A. *Chem. Eng. News* **2013**, *91* (21), 38.
3. Ainsworth, S. J.; Rovner, S. L.; Wang, L. *Chem. Eng. News* **2012**, *90* (45), 43.
4. Rovner, S. L. *Chem. Eng. News* **2012**, *90* (45), 45.
5. Gray, H. B. *Nat. Chem.* **2009**, *1*, 7.
6. Lewis, N. S. *Eng. Sci.* **2007**, *70* (2), 12.
7. The Business of Chemistry: By the Numbers. American Chemistry Council. http://www.americanchemistry.com/chemistry-industry-facts (accessed December 1, 2013).
8. *The Public Image of Chemistry*; Schummer, J., Bensaude-Vincent, B., Van Tiggelen, B., Eds.; World Scientific Publishing Co.: Singapore, 2007.
9. Smith, D. K. *Nat. Chem.* **2011**, *3*, 681.
10. Chemicals. Flash Eurobarometer 361, February 2013. TNS Political & Social. http://ec.europa.eu/public_opinion/flash/fl_361_en.pdf (accessed December 1, 2013).
11. Pavlath, A. *Chem. Eng. News* **2001**, *79* (32), 45.
12. Laszlo, P. *HYLE* **2006**, *12*, 99.
13. Francl, M. M. Don't Take Medical Advice from the New York Times Magazine, February 7, 2013. Slate. http://www.slate.com/articles/health_and_science/medical_examiner/2013/02/curing_chemophobia_don_t_buy_the_alternative_medicine_in_the_boy_with_a.html.
14. Pavlath, A. *Chem. Eng. News* **2002**, *80* (34), 51.
15. Mahaffy, P.; Ashmore, A.; Bucat, B.; Do, C.; Rosborough, M. *Pure Appl. Chem.* **2008**, *80*, 161.
16. Evans, D. A. *Chem. Int.* **2006**, *28* (4), 12.
17. Blum, D. The Rise of Chemophobia in the News? May 10, 2012. http://ksj.mit.edu/tracker/2012/05/rise-chemophobia-news.
18. CCI Solar. http://www.ccisolar.caltech.edu (accessed July 1, 2013).
19. Lewis, N. S.; Nocera, D. G. *Proc. Natl. Acad. Sci. U.S.A.* **2006**, *103*, 15729.
20. Woodhouse, M.; Parkinson, B. A. *Chem. Soc. Rev.* **2009**, *38*, 197.
21. Woodhouse, M.; Herman, G. S.; Parkinson, B. A. *Chem. Mat.* **2005**, *17*, 4318.
22. Woodhouse, M.; Parkinson, B. A. *Chem. Mat.* **2008**, *20*, 2495.
23. The SHArK Project (Solar Hydrogen Activity Research Kit). http://www.thesharkproject.org/ (accessed July 1, 2013).
24. Winkler, G. R.; Winkler, J. R. *Rev. Sci. Instrum.* **2011**, *82*, 114101.
25. Lockwood, D. *Chem. Eng. News* **2012**, *90* (46), 30.
26. Juice from Juice. http://thesolararmy.org/jfromj/ (accessed July 2, 2013).

27. The Periodic Table of Videos. http://www.periodicvideos.com/ (accessed July 2, 2013).
28. Wikipedia: WikiProject Chemistry. http://en.wikipedia.org/wiki/Wikipedia:WikiProject_Chemistry (accessed July 2, 2013).
29. Hartings, M. R.; Fahy, D. *Nat. Chem.* **2011**, *3*, 674.
30. Widener, A. *Chem. Eng. News* **2012**, *90* (50), 38.

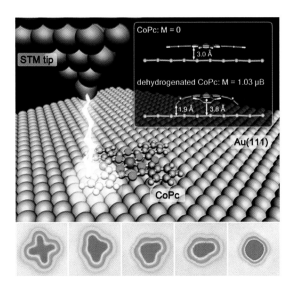

Figure 3. Illustration of manipulating single molecule magnetism. (Chapter 12)

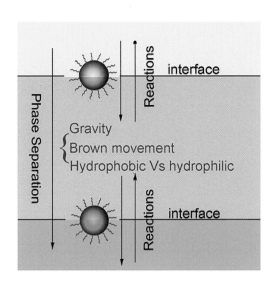

Figure 4. A novel general synthesis method for nanocrystals. (Chapter 12)

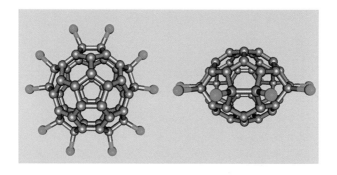

Figure 5. Schematic structure of $C_{50}Cl_{10}$. (Chapter 12)

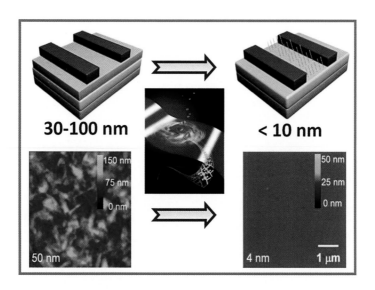

*Figure 8. Organic electronic device can be made ultrathin at nanoscale.
(Chapter 12)*

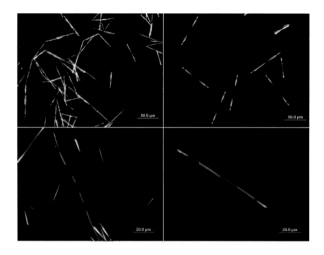

Figure 9. Yao et al. prepared organic nanoheterojucntions for optoelectronics application through molecular self-assembly. (Chapter 12)

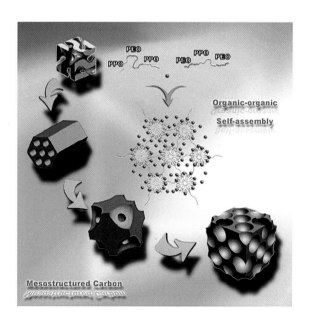

Figure 10. Mesostructured carbon. (Chapter 12)

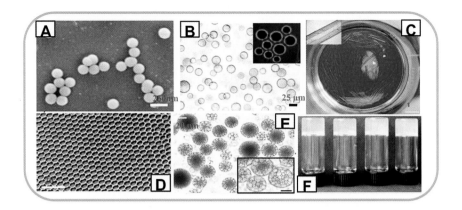

Figure 11. Self-assembly of hyperbranched polymers. (A) spherical micelles; (B) giant vesicles; (C) macroscopic multiwalled tubes; (D) honeycomb-patterned films; (E) large compound vesicles; (F) physical gels. (Chapter 12)

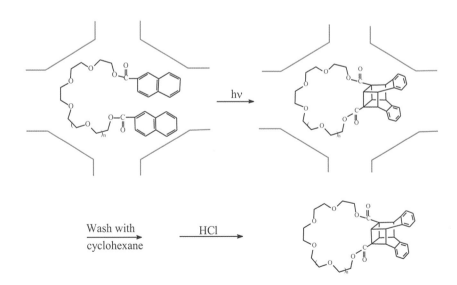

Scheme 2. Micro-reactor for photochemistry synthesis. (Chapter 12)

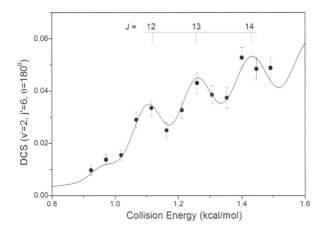

Figure 14. Experimental and theoretical DCS of the HF(v'= 2, j'= 6)product of the F($^2P_{3/2}$) + HD(j = 0) reaction in the backward scattering direction. The solid circles are experimental data; the red curve, the result of full quantum dynamics calculations convoluted with the experimental resolution and shifted 0.03 kcal/mol lower in energy. (Chapter 12)

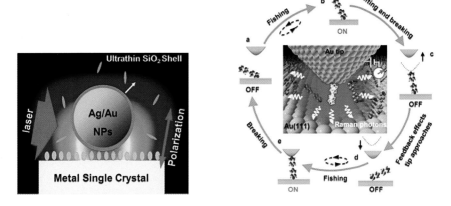

Figure 15. Left: The Principle of shell-isolated nanoparticle-enhanced Raman spectroscopy to obtain surface Raman signal on surfaces or species that do not support enhancement. Right: A schematic diagram of fishing-mode tip-enhanced Raman spectroscopy that allows the Raman and electric conductance signals of single molecules to be obtained simultaneously. (Chapter 12)

The First Homospin Single Chain Magnet

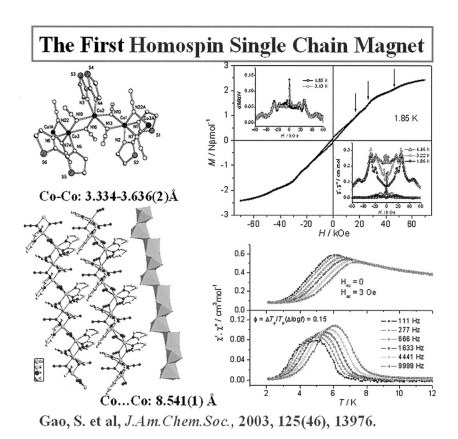

Co-Co: 3.334-3.636(2)Å

Co...Co: 8.541(1) Å

Gao, S. et al, *J.Am.Chem.Soc.*, 2003, 125(46), 13976.

Figure 16. The first homospin single chain magnet. Reproduced with permission from reference 43. Copyright 2003 American Chemical Society. (Chapter 12)

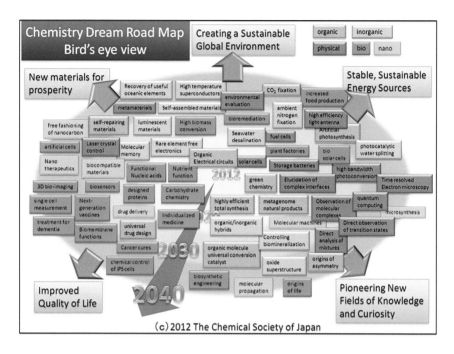

Figure 2. CSJ Chemistry Dream Road Map. (Courtesy of the CSJ). (Chapter 14)

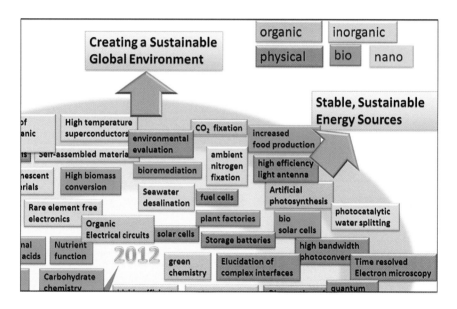

Figure 3. Close-up of the right-top part of CSJ Chemistry Dream Road Map. (Courtesy of the CSJ). (Chapter 14)

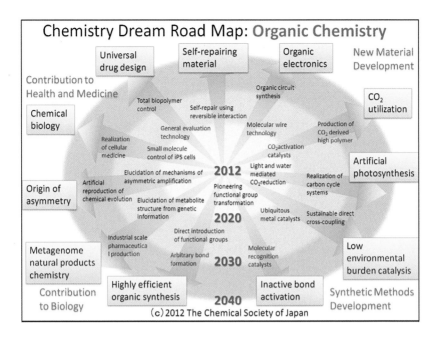

Figure 4. *CSJ Chemistry Dream Road Map: Organic Chemistry. (Courtesy of the CSJ). (Chapter 14)*

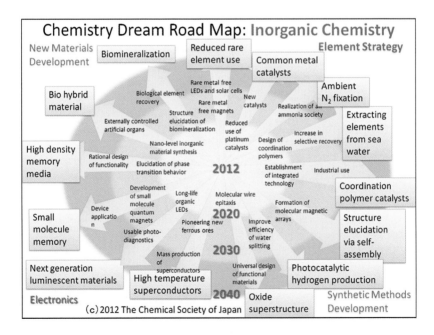

Figure 5. *CSJ Chemistry Dream Road Map: Inorganic Chemistry. (Courtesy of the CSJ). (Chapter 14)*

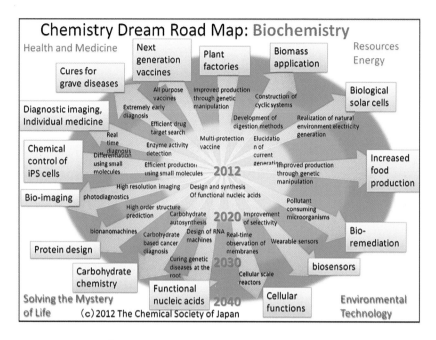

Figure 6. CSJ Chemistry Dream Road Map: Biochemistry. (Courtesy of the CSJ). (Chapter 14)

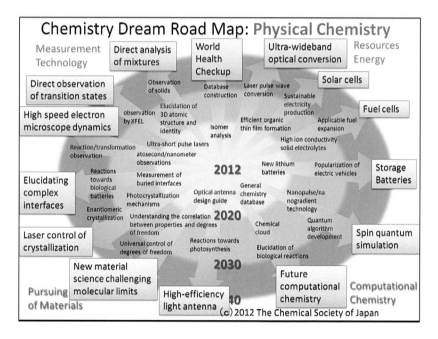

Figure 7. CSJ Chemistry Dream Road Map: Physical Chemistry. (Courtesy of the CSJ). (Chapter 14)

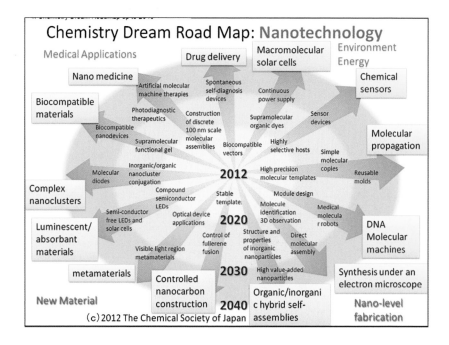

Figure 8. CSJ Chemistry Dream Road Map: Nanotechnology. (Courtesy of the CSJ). (Chapter 14)

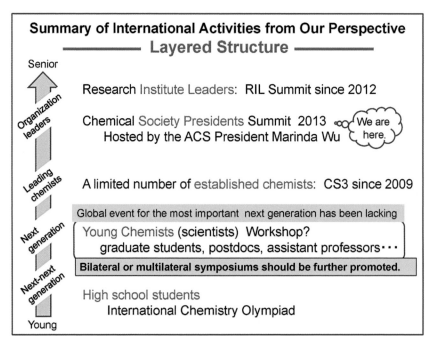

Figure 9. Summary of international activities from the author's perspective. (Chapter 14)

Chapter 6

Developing Models for International Engagement

David M. Stonner*

**Office of International and Integrative Activities,
National Science Foundation, Arlington, Virginia 22230, United States**
*E-mail: dmstonner@gmail.com

The National Science Foundation (NSF) is the nations' premier Federal agency for supporting basic research in all areas of science and engineering. As such, it has a growing interest in ensuring that our current and future science, technology, engineering and mathematics (STEM) workforce has opportunities to engage in research with colleagues from around the globe. The case for international engagement is clear: The world is becoming increasingly globalized, the research challenges we face transcend national boundaries, and research collaborations not only reduce duplication of effort, but they also enable research on a global scale that would be impossible at the national level. In recent years NSF has developed a number of approaches to increase opportunities for international engagement and to encourage students and young researchers to consider international collaboration as an integral part of their professional development.

Introduction

For over 60 years the National Science Foundation (NSF) has served to support basic research in all areas of science, technology, engineering, and mathematics (STEM). Over the years NSF has instituted many new programs and modes of supporting research to accommodate changing circumstances. In addition to support for single investigator research grants and graduate research fellowships, which have been integral to NSF since its inception, the

Foundation has instituted programs to support large instruments, develop centers of excellence, encourage multi-university partnerships, and facilitate research by small businesses, to name just a few innovations that have been developed accommodated new research opportunities.

Discussion

While NSF has always encouraged partnerships with researchers outside of the U.S., only in recent years has the need to develop more effective mechanisms for international research collaborations moved to the forefront. This interest in international collaboration has been driven less by a call from headquarters than by a demand from the research community. Not only is the sheer number of publications resulting from international collaborations growing, but these collaborations that are having a disproportionate impact on all fields of research (1). Although this trend has been growing for several decades (2), for a host of political, cultural, bureaucratic, and economic considerations, it has been difficult for research-funding agencies to develop models that proactively encourage international collaborations.

At NSF examples of dedicated mechanisms to engage international research communities include the Materials World Network and the International Collaborations in Chemistry programs. Not surprisingly, the community of U.S. chemists has been at the forefront in recognizing the value of international collaboration. An analysis of international engagement by U.S. scientists and engineers found that while only 16% of employed scientists and engineers reported international collaborations, the rate for chemists and chemical engineers was 31% and 43% respectively (2).

But other fields have not done as well in developing international connections. During his tenure as NSF Director (2010-2013), Dr. Subra Suresh, launched a number of programs to encourage international engagement across all disciplines, both for students and young researchers, as well as established researchers and centers of excellence. These experiments in international engagement not only provide concrete mechanisms for international collaboration but also send a message to the research community that NSF expects researchers to look for ways to add value to their research by engaging their counterparts globally.

Among the recent additions to NSF's catalog of internationally focused activities are Partnerships for Enhanced Engagement in Research (PEER), an alliance with U.S. Agency for International Development (USAID) to support researchers in developing countries who partner with NSF grantees; Graduate Research Opportunities Worldwide (GROW), an arrangement whereby NSF Graduate Research Fellows can conduct collaborative research for up to a year in a host country; the Belmont Forum, a mechanisms to support international partnerships on topics of global concern; and Science Across Virtual Institutions (SAVI) which matches NSF funded centers of excellence with counterpart centers abroad to jointly enhance their capabilities. Currently, more than three dozen programs across the Foundation are listed among those that highlight international collaboration (http://www.nsf.gov/od/iia/ise/index.jsp)

Despite these encouraging trends, much remains to be done. This is particularly true when it comes to providing students with opportunities for international research experiences. A recent survey of institutions of higher education found that while almost two-thirds of responding colleges and universities indicated that they offered international or global tracks or certificates for students, only 5% have international programs related to the STEM fields (*3*).

NSF has a number of programs that enable U.S. undergraduates to have research experiences abroad, including International Research Experiences for Students (IRES); international Research Experiences for Undergraduates (iREU), and Partnerships for International Research and Education (PIRE). These, however, reach only several hundred students a year and NSF will never have sufficient resources to provide more than a small fraction of the need. We are now exploring ways to develop partnerships in the private sector and non-profit sectors to provide international research experiences for a larger proportion of STEM undergraduates.

In the half century following World War II, the U.S. grew complacent with its preeminence in many areas of research. In recent years not only has the research capability of much of the world caught up, it has changed in many respects from a focus on national interests to an awareness of global concerns. The beneficiaries of greater internationalization of the U.S. research and education enterprise go well beyond the value to the participants themselves. If we are to continue to successfully compete in a global marketplace, it is absolutely imperative that we develop a STEM workforce with first-hand experience in different cultures, with different value systems, and different ways of solving problems.

References

1. Measuring Innovation: A New Perspective, 2012. OECD. http://www.oecd.org/sti/measuringinnovationanewperspective.htm.
2. *Science and Engineering Indicators 2012*; NSB 12-01; National Science Foundation, National Science Board: Arlington, VA, 2012.http://www.nsf.gov/statistics/seind12/pdf/seind12.pdf.
3. Falkenheim, J.; Kannankutty, N. *International Collaborations of Scientists and Engineers in the United States*; NSF 12-323; National Science Foundation: Arlington, VA, August 2012. http://www.nsf.gov/statistics/infbrief/nsf12323/.
4. 2012 Mapping Internationalization on U.S. Campuses. American Council on Education. http://www.acenet.edu/news-room/Pages/2012-Mapping-Internationalization-on-U-S--Campuses.aspx.

Global Opportunities from International Perspectives

Chapter 7

Innovation from Chemistry – Our Expectations of Tomorrow's Working World

Barbara R. Albert[*,1,2]

[1]Gesellschaft Deutscher Chemiker (GDCh), Varrentrappstr. 40-42,
60486 Frankfurt, Germany
[2]Technische Universität Darmstadt, Eduard Zintl-Institute of Inorganic and
Physical Chemistry, Alarich-Weiss-Str. 12, 64287 Darmstadt, Germany
*E-mail: albert@ac.chemie.tu-darmstadt.de

What are the opportunities and the challenges of tomorrow's working world – and how can we help to influence this world so that future generations will live under conditions as good as ours or even better? It makes sense to set the parameters of our societal and economical development within the world of chemistry. The framework of our chemical societies – like the American or the German Chemical Society – will help us to innovate our working world. Worldwide characteristics of the changing societies and economies are globalization, internationalization, and mobility. Huge challenges for future generations are caused by the increase of global population and the phenomenon of aging societies. This article describes how innovations for tomorrow's working world may come from chemistry! Chemical research and development are known for their strongly innovative character; chemical industry is a huge employer, and chemists have established strong chemical societies that favor diversity. Let us head towards a better balanced work environment for chemists!

Introduction

Globalization, internationalization, and mobility characterize our societies and economies. Future generations will be challenged by the increase of global population and the phenomenon of aging societies. Due to the demographic development countries like Germany, Russia, and South Korea will decrease in population while the global population continues to increase (Figure 1). Thus, in Germany we expect for the future an employment market that will be out of balance concerning supply and demand, at least in the STEM fields (STEM: science, technology, engineering, mathematics). Chemical industries contribute significantly to the economic strength of Germany and many other countries. Therefore it is important to address future opportunities and challenges that result from the combination of demographic development and globalization today.

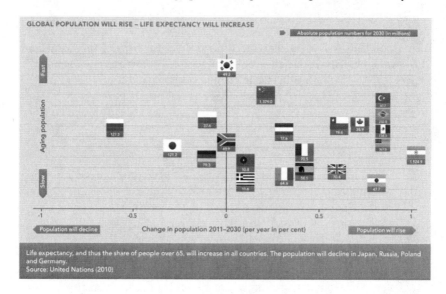

Figure 1. Change in population 2011-2030. Source: VCI (1).

The foreseeable lack of highly educated work force in Germany may be addressed on three different routes: 1) More public investement into education and research; 2) More immigration of human resources; 3) A better balanced distribution of paid work and familily time for women and men.

Not facing the demographic challenge today could result in enhanced emigration of business tomorrow – made easier in our global world.

Gesellschaft Deutscher Chemiker – German Chemical Society

With more than 30.000 members from academia, industry and other areas (26 % women) the German Chemical Society (GDCh, Figure 2) is one of the largest chemical societies in the world. GDCh is a leading member in the European Association of Chemical and Molecular Sciences (EuCheMS) which represents more than 150,000 individuals.

Figure 2. Flag with logo of GDCh, Gesellschaft Deutscher Chemiker.

GDCh maintains a constant dialogue with decision makers from academia, politics and industry and with the public for the benefit of chemistry. It supports education in chemistry at every level and promotes scientific exchange by organizing congresses and publishing journals. GDCh has a lot of publishing activities. It owns a number of internationally renowned chemistry journals, such as *Angewandte Chemie* (Impact Factor 2012: 13.734). On the other hand GDCh is a leading member in ChemPubSoc Europe, a group of sixteen European chemical societies which owns and publishes Journals such as *Chemistry – A European Journal*. Other publications include *Analytical and Bioanalytical Chemistry* and the monthly membership magazine *Nachrichten aus der Chemie*.

GDCh has 27 divisions and sections (Table 1), examples being the *Division of Analytical Chemistry* or a section which is called *Senior Expert Chemists* in which retired chemists get involved with projects like public outreach, interconnecting generations, and modern information technology for the elderly. On another – regional rather than topical – grid of structurization GDCh is organized into 61 *Local Sections* and 50 what is so called *Young Chemists Forums*. They organize lecture series and congresses with scientists from Germany and abroad, offer science lectures for the public, integrate chemistry into regional activities, and provide local contacts for journalists, teachers and students.

Table 1. Divisions and sections of GDCh – number of members (1.9.2013)

ADUC Work Association of German University Professors of Chemistry – 221

Analytical Chemistry – 2310

Applied Electrochemistry – 429

Construction Chemicals – 337

Biochemistry – 675

Equal Opportunity in Chemistry – 240

Chemistry – Information – Computer – 457

Continued on next page.

Table 1. (Continued). Divisions and sections of GDCh – number of members (1.9.2013)

Teaching in Chemistry – 1906

Chemists in Public Service – 223

Solid State Chemistry and Material Research – 848

Freelance Chemists and Owners of Free Independent Laboratories – 122

History of Chemistry – 386

Protection of Industrial Property – 240

Chemistry of Coatings – 464

Food Chemical Society – 2894

Liebig Association of Organic Chemistry – 1530

Magnetic Resonance Spectroscopy – 479

Macromolecular Chemistry – 1196

Medicinal Chemistry – 847

Sustainable Chemistry – 346

Nuclear Chemistry – 250

Photochemistry – 308

Senior Expert Chemists – 256

Environmental Chemistry and Ecotoxicology – 866

Association for Chemistry and Industry – 455

Division of Detergency and Formulations – 385

Water Chemical Society – 957

Wöhler Association for Inorganic Chemistry – 767

The board of GDCh is elected by the GDCh members and consists of seven members from industry and seven from academia or the public sector. One member of the board is elected by the divisions to represent them. The president of GDCh is elected by the board. 2012-2013, for the first time in 145 years of Chemical Societies in Germany the president is a women.

GDCh has bilateral cooperation agreements with the chemical societies of the following countries: USA, Switzerland, China, France, Spain, Republic of Korea, the Netherlands, Belgium, Poland, and the U.K. Further close contacts include the Chemical Society of Japan.

Globalization and Chemistry

Economical Relevance and Importance of Chemistry

85 % of German exports are industrial goods. The chemical industry is the third largest industrial sector in Germany follwing the automobile and electronics industries. German chemical industry exported products with a total revenue of > 150 billion € in 2011 (*1*). It has not to be emphasized that chemistry in Germany benefits from a global economy that is well interconnected. According to the *German Chemical Industry Association* (VCI, Figure 3) strong international networks are benefitial for chemistry (*2*).

Figure 3. Title of the february 2013 magazine of the German Chemical Industry Association (VCI) emphasizing the importance of global networks: "How chemistry benefits from global economy networks". Source: VCI (2).

Open markets and the abolishment of customs are considered to be not only favorable but essential for the future of chemistry. The German Chemical Industry Association strongly supports the realization of a US-EU free trade agreement, that will "... lift the economy of Europe, strengthen our economy, create jobs for Americans, for Germans, for all Europeans and create one of the largest allied markets in the world.", according to US Secretary of State John Kerry (Berlin, Feb. 26th, 2013). This quotation emphasizes two important fields of actions: 1) A global job market that is important both for industries in countries with decreasing population (like Germany) and for chemists in other countries (like USA), that struggle with a high unemployment rates in their home country; and 2) the development of global economy as it is proposed for the coming years.

According to a recent study (*3*) Europe will fall behind in the national rankings of the gross domestic product and its growth in 2030, compared to 2011 (Figure 4). The increase of the industrial production of chemicals is expected to be 4.5 % per year between 2011 and 2030, and China's contribution to that is more than 60 %. While the world production of chemically goods is expected to be centered in Asia in 2030 (Figure 5), exports of chemical goods to other European countries are expected to dominate the German chemical business.

	2011	2030	Ranking 2011	Ranking 2030	2011–2030 in per cent per year
USA	10,730	17,991	1	1	2.8 %
China	3,398	11,299	3	2	6.5 %
Japan	3,668	4,649	2	3	1.3 %
India	1,098	3,766	8	4	6.7 %
Germany	2,442	3,117	4	5	1.3 %
UK	1,880	2,649	5	6	1.8 %
France	1,811	2,543	6	7	1.8 %
Brazil	930	1,906	11	8	3.9 %
Italy	1,423	1,783	7	9	1.2 %
Canada	990	1,652	9	10	2.7 %
South Korea	859	1,555	12	11	3.2 %
Mexico	778	1,326	14	12	2.8 %
Spain	950	1,311	10	13	1.7 %
Australia	721	1,253	15	14	3.0 %
Russia	778	1,147	13	15	2.1 %

Europe will become less important to the global economy in the next 20 years. Growth in this region will be relatively low. However, China, India, Brazil, South Korea and Mexico will move up in the national rankings.

Figure 4. Growth of economy will be much lower in countries like the USA (2.8%) per year between 2011 and 2030) and Germany (1.3 %) compared to China (6.5 %), India (6.7 %), or Brazil (3.9 %). Source: VCI (1).

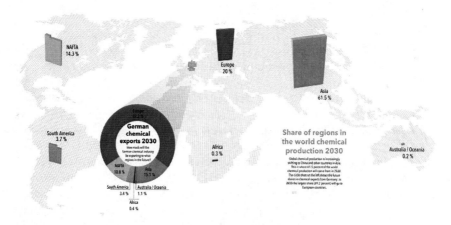

Figure 5. World production and German exports of chemical goods in 2030. Source: VCI (4).

Competitiveness in a Global World

According to Klaus Engel, the former president of the German Chemical Industry Association and chairman of the executive board of Evonik Industries, German chemical industry is planning to ensure future growth first of all by focusing on innovation. This will be achieved for example by doubling research budgets of companies until 2030, even if Germany already has a leading international role in research intensity. Research intensity, the ratio of research & development expenditures vs. production value is already almost 5,8 %, corresponding to expenditures of 8.8 billion € in 2011 (5).

But it is not only the private sector that is asked for if the degree of innovation is to be increased significantly. Innovation requires well-educated scientists and this demands to invest as much as possible into education and basic research. Several scientific societies in Germany, including GDCh, have recently expressed their sincere worry that the German goverment might neglect its extremely important task to ensure the high standard of research and education at German universities by decreasing the public funding of universities and schools. In times of very difficult financial situations of many countries this holds also for countries like the United States, Spain, Italy, and Greece, where nation-wide budget problems have led to threatening cuts in the funding of public and state universities and schools. It can be stated that the future of chemistry both in science and business is strongly coupled to the willingness of politics and societies to ensure a broard funding of top-level education and research.

Labor Shortage in Germany

The transformation of labor market is the greatest challenge for future economical growth. In Germany six million people less will be on the job market in 2030 (Figure 6). To reduce economical consequences of the demographic development in Germany the chemical industry will aim at an immigration of 200.000 highly qualified people per year and companies plan to increase the employment rate of women and to prolong working lifetime (5).

Already today the German government has passed a law for an gradual increase of retirement age. In addition to that the traditional school education was shortened by one year. Other changes that aim at a longer working lifetime will follow.

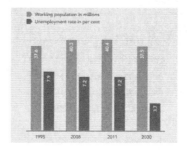

Figure 6. Labor shortage in Germany. Source: VCI (1).

Diversity in Gender and Generation

It is a well known fact that diversity enhances economical strength. Multiple studies and surveys show that economy works better when enterprises are led and managed by heterogeneous teams (6). An often-quoted example is what happened during the financial crisis in Iceland in october 2008, when the only investment fund that survived the crisis had two women at the top. While this example is by no means a representative one, it is possible to conclude from many studies and surveys published over the last ten years that the complexity of today's global economy requires decision making processes that are ruled by diverse teams, gender-wise, generation-wise and nationality-wise. Today, it is a disadvantage to choose the decision makers following the habit or the "law" of self-similarity.

In an aging society less people will have to work more and longer if the standard of life is expected to remain as high as present or even increase. With this the question of how to balance the overall work load between gender and generations becomes a very important question to address. It can be expected, that generation Y which will carry the major share of responsibility for economical and societal success in about ten years will handle the challenges of the future differently than we are able to predict. We shoud ask ourselves: Do they aim at career at any cost? Are they fully aware of the fact that today's privileges will not necessarily become a matter of course for tomorrow? If working lifetime increases, how will they interact with older generations? Do they take gender equality for granted? Do they still struggle with their vision of compatibility of employment and family?

Differences in Life Courses of Men and Women

Right now, the averaged life courses and carreers of men and women in Germany are still very different – especially when it comes to well-paid jobs with higher responsibilities. This is caused mostly by the continued incompatibility of employment and family and extended periods of part-time employment in the life courses of women.

Two examples for unequal career pathways that continue to exist: Statistics from the GDCh (7) show that female chemistry professors are still exotic in Germany (approx. 10 %) although more than 39 % of the chemistry students, including PhD students, are women; a 2011 GDCh/VAA survey of salaries (VAA: Association of Employed Academics and Executives in the Chemical Industry – Chemistry Managers) indicated that the annual income of men and women in chemistry in the upper management develops differently. In this study it appeared that the first ten years after entering professional life men and women earn approximately the same and their salary increases at the same rate. After fifteen years in business, women earn 5-8 % less than men, after 35 years the difference is almost 30 %. According to a press release from VAA in 2011 it is still the traditional role allocation that causes these differences (8). This does not seem to be a phenomen singular in Germany. Lesley Yellowlees, chemistry professor and president of the Royal Society of Chemistry in Great Britain was quoted to have attacked the "macho culture" in the UK over the lack of top women in the lab because it hurts economy (9).

In Germany, full time employment and family care are still not compatible. People that are employed part-time or unemployed obviously have little or no income and pension. One of the results is widespread poverty among the female elderly. On the other hand, a lot of men work overtime to either ensure a adequat family income or to promote their personal career, which means that they have little or no time to spend with their families. These factors have the following results: 1) Potentially valuable, well-educated women are lost for the job market;

2) Men and women have unequal career pathways and different incomes; 3) Paid work and family care are distributed non-symmetrically; 4) Increasing rates of burnout syndrome are observed.

The working world will move if we abandon traditional images of role models, if we succeed in balancing responsibilities for employment and family care, if we allow for careers without unlimited work hours (overtime) and 24 hour-availability, and if we accept different perspectives on life courses. A 2003 study from the german government showed unambigously (Figure 7) that parents who work full-time or over-time usually would prefer to work less hours and those who are not employed or work part-time would prefer to do more paid work (*10*).

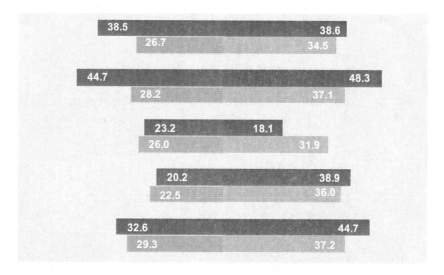

Figure 7. Hours per week (left: mothers, right: fathers). Dark colors represent the actual situation, light colors the desired work load. From top to bottom: Both parents full-time, both parents over-time, both parents part-time, one part-time/one full-time, one unemployed/one full-time. Figure drawn after BMFSF (10).

Conclusion: Tomorrow's Working World

Distributing the load of paid work and responsibilities more evenly than it is done today will help to overcome the challenges of tomorrow's labour market, at least in Germany. Also, a better understanding and collaboration amongst people from different generations and nations has to be stimulated and supported actively. Characteristics, opportunities, and challenges of tomorrow's working world comprise increasing globalization and internationalization of industry and academia. Our societies will develop growing demands on the mobility and flexibility of chemists. The tendency to replace, at least in part, long-time responsibilities by short-term projects will increase. The job market in STEM fields might be unbalanced concerning supply and demand. Working lifetime will increase and some regions will experience enhanced immigration of human beings and businesses that follow global streams, or emigration. It is obvious that compared to 2013 more women will come into employment and enter STEM careers until 2025, and women will gain more responsibility in non-family fields of work. Both men and women have to fight for a better balance of family care and employment for everybody – if work power is to be used as efficient as possible and people can live – despite increasing work life times – sustainable and holistic.

Why should innovation of the working world come from chemistry? First of all, innovation in general is known to come from chemistry. Chemical industry is striving for innovation in every context. It is at the same time a huge employer in many countries of the world. The chemical economy needs highly educated work force. Many chemists are well-paid and societally influential. They can make a difference. But, most of all one has to be aware of the fact that chemists have established strong networks forming societies like ACS, GDCh, and many more. In these societies diversity has become reality, and chemists communicate independent from generation, nationality, or gender.

"Vision 2025 – How to succed in the Global Chemistry Enterprise" (*11*) means that we unify globally our social and societal competences to shape the future of chemistry.

Acknowledgments

I would like to thank the CEO, Prof. Dr. Wolfram Koch, and his coworkers, especially Dr. G. Karger, from the head office of the German Chemical Society (GDCh), Frankfurt, Germany, for contributing to this article. Support of the VCI, especially M. Christian Buenger, is also gratefully acknowledged.

References

1. *The German Chemical Industry in 2030. A Summary of the VCI-Prognos Study*; German Chemical Industry Association: Frankfurt, Germany, December 2012, p 7 (Figure 1) and p 16 (Figure 6).
2. *Chemie Report*; German Chemical Industry Association: Frankfurt, Germany, February 2013, p 1.
3. *Die Deutsche Chemische Industrie 2030, VCI-Prognos-Studie*; German Chemical Industry Association: Frankfurt, Germany, January 2013.
4. *VCI-Factbook 06: Chemistry 2030 - Design Globalization*; German Chemical Industry Association: Frankfurt, Germany, 2013.
5. *Nachrichten aus der Chemiewirtschaft*; German Chemical Society: Frankfurt, Germany, April 2013, p 5–8.
6. McKinsey & Company. Women Matter. *McKinsey Quarterly* **2007/2008**, *1*, 2.
7. *Chemiestudiengänge in Deutschland. Statistische Daten 2012*; Eine Umfrage der GDCh zu den Chemiestudiengängen an Universitäten und Fachhochschulen: Frankfurt, Germany, June 2013.

8. VAA Press Release 6.5.2011. http://www.vaa.de/presse/presse-archiv/pressemitteilung/kind-und-karriere-fuer-frauen-immer-noch-schwer-vereinbar/.

9. The Guardian, June 2012. http://www.theguardian.com/science/012/jun/17/macho-culture-chemistry-royal-society.

10. *New Ways - Same Opportunities, First Report of Equal Opportunities*, German Government (BMFSF), 2011.

11. Wu, M. *Vision 2025: How to Succeed in the Global Chemistry Enterprise*, ACS Symposium, Spring 2013.

Chapter 8

Global Challenges Require Global Cooperation

L. Simon Sarkadi*

President, Hungarian Chemical Society, Corvinus University of Budapest, Faculty of Food Science, Department of Food Chemistry and Nutrition, 1118 Budapest, Somlói u. 14-16, Hungary
*E-mail: livia.sarkadi@uni-corvinus.hu

There are many serious problems facing the world, and chemists have the skills and knowledge to make a significant contribution to solving them. In my own area of food chemistry there are the challenges of providing sufficient high quality and nutritious food for the rapidly-increasing global population; there is the need to promote sustainability across the whole food production chain; there are the global problems of diet-related diseases such as obesity which is a significant risk factor for many adverse health conditions, and there is the need to ensure that our aging society maintains its health and well-being so as to reduce spiraling health and social costs. To address each of these global issues requires interdisciplinary research, and effective knowledge transfer to consumers, industries and policymakers. In the current economic crisis there is a need for scientists, industries, and other stakeholders around the world to work together to overcome these challenges most cost-effectively. Experienced scientists must also pass on their knowledge, contacts and experience to help train the next generation of scientists and entrepreneurs who will be involved in addressing these and other challenges.

Introduction

The Hungarian Chemical Society (HCS) was founded in 1907 (*1*). HCS represents about two thousand chemists in academia and industry. The Society has 10 Regional Bodies and 8 Workplace Groups and it supports 24 Divisions and 13 Working Parties, covering the main fields of chemistry. The Society's primary objectives include the establishment of a professional public forum for the country's chemists and provisions of indirect support for the national chemical sciences, education, and industries (including the pharmaceutical industry). One of the most important activities of the HCS is to organize national and international scientific meetings. HCS sponsors or promotes a number of international activities. As part of the International Year of Chemistry (IYC) programs, HCS and BASF established a unique online chemistry knowledge database (*2*). The website, which is currently available in 11 languages, won the *Homepage of the Year* contest in Hungary in 2011, being rated the best within the field of education. The HCS edits and supports various national and international journals and is a member of ChemPubSoc Europe and so owns a number of high-quality European journals. The HCS has established international cooperation agreements with six other National Chemical Societies and has good links with various international organizations including the American Chemical Society via the ACS Hungarian Chapter. I am very pleased to be the first female President of the Hungarian Chemical Society in more than a century, and it was a particular honor to be elected in 2011, the International Year of Chemistry.

The chemical profession has six major problems to address over the next decades: **Energy, Raw Materials, Water, Air, Health and Food**.

These are not just challenges for Hungary, of course; they are also European and global priorities. They also interact - for example, the production of healthy food requires the development of new technologies that are sustainable in their energy and water usage, the most effective exploitation of co-products (or waste products) and minimization of the pollution of the air or the wider environment. This will require global partnerships in science and technology to deliver results which can then be transferred to industry to benefit society in a competitive process.

Major Challenges Affecting the Chemistry Profession

There are many serious environmental, technological and social problems facing the world, e.g., rapidly expanding world population, limited land and water availability, climate change, energy, food and health crisis.

Chemistry, as a "central science", has a major role to play in supporting the technological and industrial sectors, and chemists and chemical engineers have the skills and knowledge to make a significant contribution to address the aforementioned problems, and to facilitating knowledge transfer and economic and social development in a sustainable way.

In 2011, the European Association for Chemical and Molecular Sciences (EuCheMS), Europe's equivalent of the ACS, prepared a Roadmap entitled *"Chemistry - Developing Solutions in a Changing World"* to highlight the central importance of chemistry to solving the pressing challenges that we, as society, face in a rapidly changing world (*3*, *4*). The role of chemistry as both an underpinning and an applied science is critical.

The following six problems are well defined by this Roadmap: Energy, Raw Materials, Water, Air, Health and Food. These are considered in turn below.

Energy

Modern life needs enormous amounts of energy to be delivered to end users as fuels, heat and electricity. Currently, the great majority of the world's primary energy supply is provided by fossil fuels (81%) and uranium minerals (5.9%). Global energy demand is predicted to double by 2050 (*5*), driven mainly by economic growth in developing countries and by the expected increase in human population from the current level of 7 billion people to more than 9 billion.

Raw Materials

Over the past 25 years, global extraction of natural resources has grown by 45%. Owing mainly to inefficient recovery and recycling, many elements are left within discarded products in landfills or are dispersed in the environment in insufficient quantities for effective reclamation. A list of critical raw materials identified at European level is shown in Table 1 (*6*).

Table 1. Critical raw materials

Antimony	Indium
Beryllium	Magnesium
Cobalt	Niobium
Fluorspar	Platinum Group Metals (PGMs) [a]
Gallium	Rare earth (REs) [b]
Germanium	Tantalum
Graphite	Tungsten

[a] PGMs: platinum, palladium, iridium, rhodium, ruthenium, osmium. [b] REs: yttrium, scandium, and lanthanides (lanthanum, cerium, praseodymium, neodymium, promethium, samarium, europium, gadolinium, terbium, dysprosium, holmium, erbium, thulium, ytterbium and lutetium).

Water

Whilst 71% of the Earth's surface is covered by water, 97% of this resource is saline. Of the remaining 3% freshwater, 90% is locked away in glaciers, polar icecaps and inaccessible groundwater. Humanity's growing needs must, therefore, be met with only 0.3% of our total water unless effective and simple desalination processes are developed and applied on a local scale. Water demand is projected to increase by 55% globally over the next four decades (5).

Air

The main environmental impacts of air pollution are acidification of the oceans, climatic changes and soil erosion through the greenhouse effect caused by carbon dioxide, methane, nitrogen oxides and ozone. Global greenhouse gas (GHG) emissions are predicted to increase by 50% (5).

Air pollution, of course, is recognized as a significant risk factor for multiple health conditions.

Health

The general indicators of human health over the past several decades show that populations are healthier, wealthier and living longer – although there is great variation and many people in developed countries do not benefit from this positive trend. However, the substantial progress in health over recent decades has been largely unequal. The nature of health problems is also changing: aging and the effects of poorly-managed urbanization and globalization all help to accelerate worldwide transmission of communicable diseases and increase the burden of chronic and non-communicable disorders. The healthcare sector will benefit greatly from new products, technologies, services and tools. The chemical sciences, in particular, have a central role to play in the fundamental research that will lead to new technologies.

Food

In my own area of food chemistry the global challenges are to provide sufficient high quality, nutritious *and affordable* food for the rapidly-increasing global population which, as soon as 2030, will have exceeded 8 billion. To match global energy and food demands with limited natural resources requires sustainability to be optimized across the whole food production chain. In addition, there are also global problems of diet-related diseases, such as obesity which is an established significant risk factor for many adverse health conditions, and there is a pressing need to ensure that our aging society maintains its health and well-being in order to reduce spiraling health and social costs, and to ensure the effectiveness of its aging labor force. Finally, it is vital that all sectors of society receive health and well-being benefits from a good diet and not just those consumers with sufficient economic means.

Opportunities for the Chemical Sciences

A greater knowledge of the nutritional content of foods is needed to understand fully the food/health interactions, which could facilitate more efficient production of foods tailored to promotion of human and animal health. The chemical sciences are keys to identifying alternative supplies of 'healthier foods' possessing an improved nutritional profile.

One of the main challenges is to produce food with less fat, salt and sugar that can be detrimental to health, while maintaining consumer perception and satisfaction from the products.

Another challenge will be to develop improved food sources and fortifying foodstuffs to combat malnutrition and to target immune system health. Even a diet that contains more energy than required can be deficient in micronutrients.

All of these challenges have to take into account the affordability of the product and will require significant input from socio-economic researchers.

Microbial contamination of food is the most common cause of health problems for consumers (either by spoilage or by adulteration) and therefore remains a critical food hygiene and safety issue. The chemical sciences can support the European Food Safety Authority (7) in securing Europe's food safety.

International Organizations

The international organizations take a variety of forms and address a number of global challenges, notably food security, agriculture, health, energy, and climate change.

Food and Health Research in Europe (FHARE)

FHARE (8) aims to increase the structuring of food and health research and support cooperation towards the European Research Area (ERA).

In April 2012, FAHRE launched its proposals for food and health research in Europe. FAHRE proposes more research on how to achieve healthier eating, e.g., through changing food production, changing behaviors, and showing the impacts of better governmental policies and regulation. It also proposes improvements in the organization of food and health research in Europe. There should be better links between food research and medical research, and more use of social sciences to determine effective interventions.

Joint Programming Initiatives (JPI)

Another example of the efforts of the European Union to tackle grand challenges is through joint programming (JP).

A Healthy Diet for a Healthy Life (HDHL) provides a roadmap for harmonized and structured research efforts in the area of food, nutrition, health and physical activity (9). The JPI "A Healthy Diet for a Healthy Life" will launch three Joint Actions in the following areas in the near future:

- **Determinants of Diet and Physical Activity** (DEDIPAC)
  Research will include studies which aim to improve understanding of the different biological, psychological and socio-cultural factors that impact on health, and how they interact

- **Roadmap Initiative for Biomarkers for Nutritional/Health Claims**
  The objective is to develop guidelines for a dossier for health claims.
- **European Nutrition Phenotype Data Sharing Initiative**
  The objective is to provide the highest level of standardization of all phenotypic information of study subjects with regard to diet, physical activity levels and all biological, clinical and physiological measurements that define human body responses in health and disease states.

European Technology Platform Food for Life (ETP F4L)

European Technology Platforms (ETP) are industry-led, public/private partnerships encouraged by the European Commission to drive innovation and unite stakeholder communities in reaching strategic research objectives of key European industry sectors (*10*). The main goals of the ETPs are to strengthen the European innovation process, improve knowledge transfer and stimulate European competitiveness across the food chain.

ETP Food for Life identified the following areas of scientific challenge (*11*):

- Improved Health, Well-being and Longevity
- Safe Foods that Consumers Can Trust
- Sustainable and Ethical Production
- Food Processing, Packing and Quality
- Food Chain Management
- Communication, Training and Technology Transfer

European Institute of Innovation and Technology (EIT)

The 8th Framework Programme (now called Horizon 2020) will run between 2014 and 2020. Its priorities include addressing Global Challenges **integrating the Education, Research and Innovation sectors** (Note: ETP Food for Life only brings together the latter two) within the European Institute of Innovation and Technology and promoting collaboration and cooperation between the EU and other regions of the world (*12*).

EIT governance structure is based on a Governing Board and Knowledge and Innovation Communities (KICs). The three existing KICs are: climate change mitigation (Climate-KIC) (*13*), information and communication technologies (EIT ICT Labs) (*14*) and sustainable energy (KIC InnoEnergy) (*15*).

Conclusions

To address each of these global issues, chemists need to be able to work together with other professions and specializations, including the social sciences, and with industry. The fundamental element of measurements of the chemical sciences offers chemists an immediate entry into such areas of cooperation.

It is also vital to support and encourage the next generation of chemical and molecular sciences and to provide them with contacts and networks which are necessary to optimize the impact of the results they produce.

The above described priorities are extremely challenging and demand that countries and regions work together to deliver reliable information to stakeholder communities.

Acknowledgments

Thanks are due to Dr Marinda Li Wu, President of the American Chemical Society for inviting me to the Presidential Symposium on "Vision 2025: How to Succeed in the Global Chemistry Enterprise" in New Orleans on April 8-9, 2013 and giving me the opportunity to share my thoughts on global challenges facing the chemical world.

References

1. Hungarian Chemical Society. http://www.mke.org.hu (accessed March 2013).
2. chemgeneration.com. http://www.chemgeneration.com (accessed March 2013).
3. European Association for Chemical and Molecular Sciences (EuCheMS). http://www.euchems.org (accessed March 2013).
4. *Chemistry - Developing Solutions in a Changing World*; The European Association for Chemical and Molecular Sciences (EuCheMS): Brussels, Belgium, 2011.
5. *Environmental Outlook 2050*; The Organization for Economic Co-operation and Development (OECD): Paris, 2012.
6. Critical Raw Materials for the EU, 2010. European Commission. http://ec.europa.eu/enterprise/policies/raw-materials/files/docs/report-b_en.pdf.
7. European Food Safety Authority. http://www.efsa.europa.eu/ (accessed March 2013).
8. Food and Health Research in Europe, 2010. FAHRE. http://www2.spi.pt/fahre/reports/research_needs_synthesis.pdf.
9. Healthy Diet for a Healthy Life. JPI. http://www.healthydietforhealthylife.eu/index.php/joint-actions (accessed March 2013).
10. European Technology Platform, Food for Life. http://etp.fooddrinkeurope.eu/asp/index.asp (accessed March 2013).
11. *Strategic Research and Innovation Agenda 2013-2020 and Beyond*; European Technology Platform, Food for Life: Brussels, Belgium, 2012.
12. European Institute of Innovation and Technology. http://eit.europa.eu/ (accessed March 2013).
13. EIT Climate-KIC. http://www.climate-kic.org/ (accessed March 2013).
14. EIT ICT Labs. http://www.eitictlabs.eu/ (accessed March 2013).
15. KIC InnoEnergy. http://www.kic-innoenergy.com/ (accessed March 2013).

Chapter 9

Chemical Education – A Key Factor in Facing the Challenges of the Future

Sorin I. Rosca*,1 and Cristina Todasca2

1President of the Romanian Chemical Society,
Department of Organic Chemistry, University Politehnica of Bucharest,
313 Splaiul Independentei, Bucharest, Romania
2Chair of Young Chemists, Division of the Romanian Chemical Society,
Department of Organic Chemistry, University Politehnica of Bucharest,
313 Splaiul Independentei, Bucharest, Romania
*E-mail: si_rosca@yahoo.com

With the aim of finding out if Romanian chemists are properly prepared to participate in the general effort to address the enormous challenges of the future, an analysis was performed by the Romanian Chemical Society (SChR). It emphasized some peculiar aspects of weaknesses in chemical education at various levels due to its commune origin. Among these are the chemists' difficulty in communicating with ordinary people due to both the esoteric scientific language and limited chemical knowledge of the general public; insufficient in-depth training of chemists in other sciences – physics, biology, mathematics – thus hindering multidisciplinary collaboration; relatively low interest of young people to study and teach chemistry; and the elimination of quantitative criteria for evaluation of scientific activity at the cost of creativity and applicability of scientific results. SChR has initiated specific activities designed to ameliorate the aforementioned weaknesses. For instance, in order to increase the attractiveness of chemistry for young people, local branches, Chemical Education and Young Chemists sections of SChR in collaboration with universities have organized interesting and challenging scientific contests for pre-college students. Topics like "Chemistry, friend or enemy?", "How is prepared?", "Experimentum", "Science

Fair", "Human. Environment. Pollution" are obvious in their content and expectations.

Introduction

At the beginning of September 1913, at the BASF plant in Oppau, Germany, the first industrial installation for the manufacture of ammonia was inaugurated through synthesis from its elements. The plant had a capacity of 30-40 tons/day, i.e., less than 10000 tons/year (*1*). Today hundreds of plants worldwide manufacture more than 100 million tons of nitrogen fertilizers per year. What determined such a fabulous increase in production?

A complex socio-economical study conducted by specialists of the University of Manitoba, Canada (*2*) shows that the world development till now, when the population reached 7 billion, would not have been possible without the existence of the Haber-Bosch process for the manufacture of ammonia. This is maybe the most objective evaluation of the role played by chemistry in the development of modern civilization, because it is not based on an uncertain prognosis, but on an analysis of real historical facts.

Looking towards the future, the role of chemistry can be clearly foreseen. For the start we should consider the prognosis of a population increase of 2 billion till the middle of the century, more than the entire world population in 1900, which consisted of 1.6 billion people. It should be mentioned also that these figures do not truly reflect the real needs of necessary subsistence. One should consider the necessity to eradicate malnutrition which still exists in large parts of the world as well as the increases in consumer demands in the emerging countries, in order to appreciate more realistically that the food production has to double by the middle of the 21st century.

Chemistry is expected to play a big role and not just to provide synthetic fertilizers. It is expected to play a role in the manufacturing of textiles for apparel and other uses, in new methods for producing and storing energy, and in the production of drugs which will increase life expectancy (*3*). Thus, scientific discoveries and invention of new technologies are needed, which in turn require scientific research.

In order to achieve these goals the starting point is the availability of a highly educated, intelligent, work-driven and highly competent human resources. Education is necessary both for the formation of a research elite capable of advancing chemical knowledge and for the general education of the entire society. The general public should be aware and understand the challenges, objectives and needs of chemistry researchers.

Romanian Chemical Society (SChR)

Brief History

The Romanian Chemical Society was founded in 1919, by Royal Decree (*4*). This remains the official date, although the society is a continuation in time of other much older professional scientific organizations of Romanian chemists.

Indeed, since 1890 "The Romanian Society for Sciences" existed in Bucharest, which had a powerful Chemistry Section, and since the beginning of the 20th century the "Society for Sciences" existed in Iasi. Both societies were founded by eminent chemists: Constantin Istrati in Bucharest and Petru Poni in Iasi. Acknowledging the advantages offered by specifically mentioning "chemistry" in the title of the newly formed society in 1919 (for example, the possibility for this society to be affiliated with IUPAC), the chemistry sections of the societies in Iasi and Bucharest joined together to form the Romanian Chemical Society, bringing their expertise and a remarkable international reputation reflected by the fact that some of the most important personalities of the time were honorary members of the society in Bucharest. Among them may be mentioned: Marcellin Berthelot, Charles Friedel, Victor Grignard, Paul Sabatier from France, Adolf von Bayer, Emil Fischer, August Wilhelm von Hofmann, Friedrich Kekule from Germany, Dimitri Mendeleev from Russia, and Stanislao Cannizzaro from Italy.

Figure 1. The logo of SChR (left). "Petru Poni" Medal of the Romanian Chemical Society (right). Courtesy of Romanian Chemical Society.

The scientific societies in the Romanian territory had a much older history. Probably beginning in the 1830's-1840's in Iasi, Bucharest and Sibiu, in every historic province of Romania existed societies for medicine and natural sciences which had notable chemistry sections. These societies also had as honorary members great personalities of chemistry as Jakob Berzelius, as well as naturalists such as Alexander von Humboldt and Charles Darwin.

As we are speaking of the activities of SChR at its beginnings, we will mention as its permanent objectives the organization of scientific congresses and the debate of chemical problems at the international level. Affiliated to IUPAC immediately after its creation in 1921, SChR received almost immediately in 1925, the mission to organize the 6th IUPAC congress, the mission well accomplished according to the documents of the time (5). The National Congresses were held regularly, the 5th being held in 1936.

The Society published the "Bulletin of the Chemical Society", the "Bulletin of the Pure and Applied Chemistry" (a scientific journal presenting original results published in languages for international circulation), as well as the "Chemical Bibliography in Romania" (a journal which included original work in the manner according to international review standards).

Then the war years came followed by a postwar era in which the Society had a long period of hibernation. Without having been dissolved, it did not exist *de facto*, the authorities of the time being not in favor of such organizations. Only in 1992 was SChR resuscitated, goals were set, and reorganization was achieved. International contacts were also developed.

Present Activity of SChR

Today SChR (current logo and honorary medal in Figure 1) has approximately 3000 members including many affiliated members (young and retired members). The society is organized in 20 local branches across the country, based in cities important for education and chemical industry (Figure 2 left). A parallel organization in 7 sections is based on professional specialization, which, according to the SChR statute and the European model, covers the main domains of chemistry to which are added a "Section for Chemical Education" as well as an innovative "Young Chemists Section".

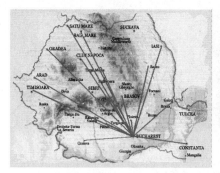 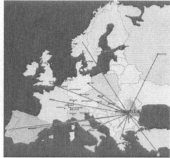

Figure 2. Local branches of SChR (left). European SChR relations within EuCheMS (right). Courtesy of Romanian Chemical Society.

A key activity of SChR is to organize scientific meetings. Regional cooperation with the chemical societies from Albania, Bulgaria, Cyprus, Greece, Macedonia, Montenegro and Serbia has led to the organization of "International Conference of the Chemical Societies of the South Eastern European Countries" (Figure 3 left). In cooperation with the universities of Bacău (Romania) and Orleans (France), the "Franco-Romanian Colloque of Applied Chemistry" (COFrRoCA) has been organized (Figure 3 right). In cooperation with the University "Politehnica" Bucharest, "Ovidius" University Constanţa, "Petru Poni" Institute Iasi, international conferences have been organized periodically at a high scientific level.

The present publications of SChR include "Revista de Chimie" (a scientific journal publishing original results, with a constantly rising ISI impact factor), the "Bulletin of the Romanian Chemical Society" (an informational journal on chemistry-related activities in Romania), "Chimia" (a publication for undergraduates), as well as a series of publications of the local sections of SChR: "ProChimia", "Universul Chimiei", and "ChimMax".

The local sections of SChR together with the universities have as an important task the attraction of young people to the study of chemistry. Some selected activities directed towards this goal are presented below.

Figure 3. International Conference of the Chemical Societies of the South Eastern European Countries (left). Franco-Romanian Symposium on Applied Chemistry (CoFrRoCA) (right). Courtesy of Romanian Chemical Society.

Following the before-the-war tradition after 1992, SChR promoted international cooperation, and as a result of its actions SChR was chosen to organize the General Assembly of the European Chemical Societies in Bucharest in 2004. At the general assembly meeting hosted by the Romanian Chemical Society and held in the historic Parliament building, the member societies approved an amended statute and a new name (Figure 4). The European Association for Chemical and Molecular Sciences (EuCheMS) was founded on this occasion. In the framework of this association (EuCheMS) or through bilateral agreements, SChR has cooperative relations practically with all European countries (Figure 2 right).

EuCheMS takes over from FECS

NEWS RELEASE

EuCheMS, the new European Association for Chemical and Molecular Sciences will take over the role and responsibilities of the former Federation of European Chemical Societies (FECS).

At the General Assembly meeting hosted by the Romanian Chemical Society and held in the historic Parliament building in Bucharest, the member societies approved an amended constitution and the new name. Following due process in the coming months, EuCheMS will become an 'Association Internationale Sans But Lucratif' (not-for-profit organisation) in Belgium.

EuCheMS will build on its 30 years of history and will aim to provide a more professional level of support for the needs of its 50 member societies across 36 countries throughout Europe. A funding base will enable EuCheMS to modernise its approach and develop meaningful support for chemical and molecular sciences in the 21st century.

Figure 4. Part of the announcement regarding the formation of EuCheMS.

Of great interest is the cooperation with the American Chemical Society (ACS). Paramount in the development of this cooperation was the visit, in October 2012, of Dr. Marinda Li Wu, president-elect of the ACS, who contributed decisively to a better understanding and a main pathway to cooperation. The conditions for this cooperation has been optimized in the future through the creation of the new ACS International Chapter for Romania.

Problems in the Professional Life of Romanian Chemists

The National Council of SChR examines in semestrial meetings the activities of the society both in the main sections (education, publications, scientific research, internal and international professional networking) and in the global setting. On these occasions, problems encountered by the Romanian researchers and by SChR are identified, so that appropriate remedies may be prescribed.

Decrease in the Interest of the Younger Generation To Study Chemistry

One of the problems, constantly observed lately, is the progressive decrease in interest of the younger generation to study chemistry or to choose a profession in this field. If 20-25 years ago 3-4 candidates competed for a place in the Faculties of pure or applied chemistry, today the same faculties have hardly one candidate per vacancy. Sadly the decrease in the number of candidates is not compensated by an increase of their motivation so that the average professional quality and quantity of the students is decreasing. Of course there are many causes for this situation. An important cause is the clear decline in the chemical industry in Romania due to the change in national economy and to the conjunctional effects of the economic crisis. Production has been drastically reduced in areas of chemical industry such as fertilizers, acids and bases, synthetic rubber, synthetic fibers, cement, calcium carbide and others – the same areas in which Romania was known to be among the top 10 world producers in 1985 (6). It is easy to understand why the number of jobs decreased in the research laboratories and in the institutes of industrial design, which used to attract many competent young people in the past.

Another reason for the decreased interest in choosing a career in chemistry is the imperfect state of basic education in chemistry as pointed out below.

Imperfections in Basic Education in Chemistry

In the undergraduate education system in Romania, chemistry is often regarded by students as an uninteresting discipline. Paradoxically a field of science, which is so closely related to the needs of a high quality of life, is too often suffocated by memorization of a multitude of classifications and schemes of reactions. Instead of an education favoring creativity, a high density of rote knowledge is preferred, part of which does not provide a general education and probably will never be useful in life. Even the curricula for high-schools with literature and art concentration or vocational schools contain too much rote information regarding chemistry. Much more useful would be knowledge about applications and uses of substances, and, when appropriate, dangers and safe modalities of lab manipulations.

Probably our high school education suffers mainly due to the lack of an experimental curriculum. This type of activity which is so stimulating for the curiosity of young people and so efficient for the accumulation of solid knowledge is sadly very rare, due to the decrease in the number of hours dedicated to chemistry and to the scarcity of experimentation materials.

In higher education scientific research is the most efficient method of professional training, but the regulations regarding the evaluation of PhD students and young researchers tend to accentuate the quantitative aspect (e.g., number of publications) and not the qualitative aspect (which is more difficult to express numerically). This can lead to undesirable consequences, such as the decrease in the satisfaction in the research work as well as a proliferation of mediocre work. In time this leads to the decrease or even the lack of passion in the research work and diminished attractions to chemistry.

Deficiencies in the Communication with the Public and the Authorities

Communication between the Romanian chemists and the public is far from ideal in terms of efficiency. On the contrary, chemists are concerned with their professional problems. They speak an esoteric language, ignoring the necessity to "translate" the information. In so doing, they gradually lose the understanding and support of the public.

A superficially informed public accepts more easily a deformed presentation of chemistry in which the benefits are diminished and the negative effects are sensationally presented (e.g., exaggerating the dangers of synthesized products and exhibiting unreasonable aversion towards them).

It is no wonder then that in some cases justifiable requests, such as to stop reducing the number of hours devoted to the study of chemistry in schools or to better equip laboratories in schools, are not supported by the public. We hope that the state authorities will take logical actions in order to minimize discrepancies between the scientific research needs and the public funding allocated. (Today in Romania, according to SChR statistics, chemistry accounts for 24% of scientific publications and 44% of the citations of original results but gets only 10% of the public funding.)

According to the data above, the main weaknesses in the chemistry profession stem from the deficiencies in the chemical education. This refers to both the education of research elites which will become professionals in chemistry and to the chemical education of the general public.

In Romania basic education is directed by the state authorities, the Ministry of National Education being in charge of the curriculum, the organization, and the control of undergraduate education. According to its statute, SChR assumes the role of an active social partner in interacting with the political system and advocating for education in chemistry, as well as the role to initiate and perform activities with respect to the promotion of chemistry. In the following we will illustrate some of these activities.

Program of Activities: "2008 – A Year for Promotion of Chemistry in Romania"

In 2007 the National Council of SChR announced and one year later implemented a program of activities with the above title. The program aimed to achieve in one year a number of important actions directed towards the demonstration of the role and the importance of chemistry in the modern society. This was not conceived as a one-year program but as a campaign with many initiatives, the most effective ones hopefully widely disseminated and made permanent.

The activity program was designed according to the fundamental principle in the SChR statute (7) which required the Society "to promote chemistry in all aspects." It was also consistent with EuCheMS vision regarding the communicating chemistry to the public: "It is not sufficient to undertake activities; there is a real need to disseminate the information and we must promote these and publicize them to our own societies and their members and to other bodies, National and European institutions".

Some of the main goals of the program were:

- To promote in the public opinion a fair understanding and attitude with respect to chemistry;
- To intensify actions for the orientation of young people to study and practice chemistry as a profession
- To contribute to the improvement of the quality of education and scientific research in chemistry;
- To increase the visibility of SChR at national and international levels.

The program "2008- a year for promotion of chemistry in Romania" proved to be very effective and created an effervescence of ideas, which generated many initiatives proposed by local sections, many of which have become generally accepted.

It must be mentioned that in the work of SChR and probably in that of many sister societies there is a need to have an agreement of the ideas, goals and methods used for the promotion of chemistry at an international level. The International Year of Chemistry 2011 was positively received in Romania: the actions initiated benefited not only from the authority conferred by SChR and EuCheMS, but also from the communion of ideas regarding the challenges, goals and requirements of chemistry. More recently, in April 2013, ACS organized at their Spring National Meeting, the symposium "Vision 2025: How to Succeed in the Global Chemistry Enterprise", an excellent initiative on global cooperation of the sister chemical societies brought by Dr. Marinda Li Wu, the president of ACS. It is interesting to see that, regardless of size and experience, the chemical societies from different continents experience essentially similar problems and consequently their visions are worth sharing. At the same time the feeling of solidarity is comforting, thereby encouraging bolder and safer actions.

In the following will be presented some successful activities of SChR. It is not an exhaustive report but consists only of some specific examples regarding the stimulation of young people's interest in chemistry.

Activities of SChR To Attract the Young Generation To Study Chemistry and To Choose It as a Profession

Once adopted as a goal, interest in attracting the young generation towards chemistry became important for the local sections which proposed the initiatives. They also facilitated this process, usually by forging partnerships with universities, high schools, authorities and sponsors.

Although differing by topic, dimension and organization, these activities have as a common feature the idea of offering to the young generation the possibility to meet and to discuss chemistry beyond the curricula in creative ways.

Figure 5. Poster on the national contest, "Chemistry, friend or enemy?" (left). A presentation session in the "Chemistry, friend or enemy?" contest at Bucharest University (right). Courtesy of Romanian Chemical Society.

As an example, "Chemistry, friend or enemy?" (Figure 5) was an activity conceived as a national symposium, in which students presented reports regarding the uses and dangers of different substances and chemical phenomena. Started with presentations where the teachers proposed the topics and suggested the bibliographies, the symposium evolved to comprise problems allowing more substantial personal activity, including experimental work by the students. As an example, in the summary of the report: "Lake Ciuperca friend or enemy?", the authors mentioned, "In order to establish the ecological status of Lake Ciuperca we selected a set of indicators determined by different methods of analysis: pH and electric conductivity measured by potentiometric methods, dissolved oxygen and chlorides by titrimetric methods, nitrogen from nitrites, and phosphorus from orthophosphates by UV-VIS molecular absorption spectrometry. From all indicators Lake Ciuperca should be classified in quality class II" (*8*).

In the same manner such topics as "biofuel obtained from vegetable oils", "determinations of sugars and vitamin C from fruits", and "chemistry in colors – inorganic pigments" were presented.

Figure 6. High school students demonstrating an experiment in front of the jury at the "Experimentum" contest. Courtesy of Romanian Chemical Society.

The constant reorientation to increase emphasis on experimental activities in this symposium was not unexpected, nor was it isolated. It came from the tendency to fill, even partially, the gaps in experimental instruction of the undergraduate studies, and it followed the recommendations of SChR. Many of the activities promoted by the local sections and the sections of the society belonged to this type of initiative. The contest "Experimentum" (Figure 6) organized by the Arad local branch in western Romania consisted of experiments performed by teams of 2-4 students. Experiments were prepared under the direction of a chemistry teacher, a report was presented, and a practical demonstration performed in front of a jury, the best presentations being awarded prizes and diplomas from SChR. The contest was attractive due to the interesting topics (e.g., isolation of colorants from natural sources and their use as a pH indicator, invisible inks, reactions with extraordinary effects, electrolyses, and fabrication of electric batteries) but also due to the satisfaction of working in a team and to the honor of representing one's school and city.

Recently a more involved contest was proposed and put to practice. At the initiative of BASF-Romania, the educational project "Chain Reactions" was organized in cooperation with SChR and high schools in Romania. This consisted of advertisements and putting into practice of complex projects involving a series of physical and chemical phenomena by teams of high school students under the supervision of chemistry and physics teachers.

Each phenomenon was started automatically through the effect of the preceding one, so that after the start, the ensemble worked without human intervention. The success of the project was beyond expectations as regarding the number of participants, the ingenuity, and the depth of knowledge of the students. The contest will most probably become a tradition in Romania, and it is a very good example for future cooperation of SChR with representatives of the chemical industry.

Figure 7. Products made by students in the "How is prepared?" competition (left). A teenage style poster for the "How is prepared?" competition. Courtesy of Romanian Chemical Society.

Figure 8. Exchange of ideas in the "Students for Pupils" activity (left). Students demonstrating an experiment to pupils in the "Students for Pupils" activity (right). Courtesy of Romanian Chemical Society.

The idea of encouraging experimental activity was also behind the project "How is prepared?" (Figure 7) which involved a series of presentations followed by experimental demonstrations on the preparation of useful substances and materials, such as a perfumed soap, an after-shave lotion (prepared by extraction from natural materials), and a flavor or a dye obtained by synthesis. It must be noted that the whole project from its conception to organization and evaluation belonged to the "Young Chemists" section of SChR, and its popularity and prestige grew with every edition. The same "Young Chemists" section also initiated two other actions: "Students for Students" and "Students for Pupils" (Figure 8). Meetings of college students with high school students were thus a very effective way to attract teenagers to chemistry. The credibility of the college students in front of the high school students, the emulation which they created in the mind of students only a couple of years younger was very high due to the similarity in the way of thinking, their sincerity and their friendly attitude.

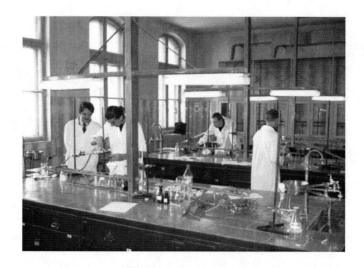

Figure 9. Practical work in the "Nenitescu Contest". Courtesy of Romanian Chemical Society.

One of the most common forms of encouraging the study of chemistry, valid practically worldwide, is the participation to competitions like the International Chemistry Olympiads. SChR participates both in the preparation of the National Chemistry Olympiads (which is organized by the Ministry of National Education) as well as in initiating and supporting of regional competitions. An important role is played by the "Nenitescu Contest" which is named after the most famous Romanian organic chemist. Organized according to the rules of International Olympiads and having a high degree of difficulty, the "Nenitescu Contest" is a demanding test and excellent training for the Olympiads. This is proved by the fact that, almost without exception for almost two decades, the Romanian laureates of the International Olympiads were chosen from the laureates of the "Nenitescu Contest" (Figure 9).

The series of activities conceived to attract young people to chemistry will be continued. We will only mention that these activities can have very diverse forms, beyond the conventional ones. One example was revealed in the contest "Human. Environment. Pollution" (Figure 10). Beyond the conventional form of organization (presentations, experimental work) this contest included also a poster session which had a positive educational effect beyond expectations.

It was difficult to anticipate the large emotional impact of an ecologic idea when it was not expressed in usual scientific terms, but this happened in the case of a drawing with the title "My friend, Terra is crying!"

Beyond expectations was also the impact of the activities organized for the youngest pupils and even for the children in kindergarten. Brought in to play the games of "little chemists", they showed a precocious interest and enthusiasm, which may hopefully influence their future attitude towards chemistry (Figure 11).

Figure 10. "My friend, Terra, is crying", a poster in the contest "Human. Environment. Pollution." (left). A model of an ecological house made by a pupils' team in the contest "Human. Environment. Pollution" (right). Courtesy of Romanian Chemical Society.

Figure 11. Primary school pupils answering questions on chemistry (left). Kindergarten children playing as "little chemists" (right). Courtesy of Romanian Chemical Society.

Acknowledgments

The authors express their special thanks for the help in preparation of the manuscript to Dr. A. Hartopeanu, Dr. R. Atasiei and Dr. H. N. Cheng.

References

1. Abelshauser, W.; von Hippel, W.; Johnson, J. A.; Stokes, R. G. *German Industry and Global Enterprise BASF: The History of a Company*; Cambridge University Press: Cambridge, U.K., 2004.
2. Smil, V. *Enriching the Earth: Fritz Haber, Karl Bosch, and the Transformation of World Food Production*; MIT Press: Cambridge, MA, 2004.
3. *Chemistry - Developing Solutions in a Changing World*; European Association for Chemical and Molecular Sciences: Brussel, 2011.
4. Decret Regal no.2351, Monitorul Oficial no.9, 5.06.1921, Bucuresti, Romania.
5. Gane, G., Ed.; *Guide de la VI-ème Conference Internationale de la Chimie Pure et Appliquée*, Bucuresti, 1925.
6. *Anuarul Statistic al Romaniei*; Institutul National de Statistic: Bucuresti, 1986.
7. Statutul SChR, *Buletinul Societatii de Chimie din Romania* **1993**, *2*, 1–7.
8. Arhip, A. D.; Malihin, M. D.; Parau, S. *Simpozionul 'Chimia prieten sau dusman"*; Universitatea Bucuresti: Bucuresti, Romania, 2013.

Chapter 10

Mexico Is Working Towards a Global Chemistry Community

Cecilia Anaya Berrios*

President, Sociedad Química de México, Mexico D.F., 03940, México, and
Academic Vice President, Universidad de las Americas Puebla, UDLPA,
Cholula, Puebla, 72810, México
*E-mail: cecilia.anaya@udlap.mx

We need to work on the education of our kids and young people to improve their knowledge of the sciences, particularly chemistry. The general public needs to be aware of the risks and the benefits that are involved in the responsible use of chemistry. In Mexico there are now also more undergraduate and graduate programs in chemistry and related fields that have been accredited with high standards. This has increased the number of students involved in science and technology. In addition, the chemical industry and universities in Mexico have been working in a collaborative way to develop new technologies and new products. The industry and the government provide funding to support research projects. Moreover, there have been international agreements to increase collaborations in education and research among universities and governments, but we need to enhance our efforts on this front in order to promote a global chemistry community.

The Mexican Chemical Society (Sociedad Mexicana de Química, SQM) was founded in 1959 and its mission is "to contribute to the growth of chemical sciences, technology and industry as well as the dissemination of chemical knowledge and the awareness of general public." Its vision is: "Keep updated our members with respect to scientific, technological and industrial chemical demands."

The goals of the Mexican Chemical Society are five:

1. Development of outreach activities for general public.
2. Fostering of information exchange and scientific research among the members.
3. Promotion of events, programs, products and services for our members.
4. Cooperative activities with other chemical societies in a global perspective.
5. Organization of activities with a focus on sustainable development and innovation

In this regard, the Mexican Chemical Society have been working through several activities to enhance public awareness of the benefits of chemistry. A relevant activity was the construction of a gigantic Periodic Table (Figure 1) during the festivities of the International Year of Chemistry in 2011. The success of this initiative was due to the active participation of national higher education institutions, i.e., student, faculty and administrators. The dimension of each element is a cubic meter when it is set, and it is only about 20 square centimeters when it is folded. This has allowed us to transport it easily. It has been exhibited at most of the institutions that have collaborated in its preparation and at public places where people could spent a day learning more about each element and its applications.

To foster information exchange and scientific research among the members, the Mexican Chemical Society organizes meetings, workshops and conferences on a regular basis. The most relevant is the annual national meeting. Last year it was co-organized with the Federation of Latin American Chemical Associations (Federación Latinoamericana de Asociaciones Químicas, FLAQ).

We also have been working with other chemical societies around the world. Of course we would like to sign Memoranda of Understanding (MOU) with them to begin more fruitful interactions in order to benefit our members. It is important to mention the support of the American Chemical Society (ACS) and the International Union of Pure and Applied Chemistry in our Latin American Chemical Conference 2012 (Conferencia Latinoamericana de Química, CLAQ 2012), which took place at Cancun, Q.R. Mexico.

In addition, the chemical industry, chemical societies, and universities in Mexico have been working in a collaborative way to develop new technologies and new products, and carrying out other interactions. We have also established in 2012 a prize for young researchers in Green Chemistry, which is sponsored by BASF-UDLAP

Working together, we aim to enhance our efforts towards the promotion of a global chemistry community.

Figure 1. Display of the Period Table constructed during the festivities of the International Year of Chemistry in 2011.

Chapter 11

The Brazilian Chemical Society (SBQ) and the Global Chemistry Enterprise: Building a Sustainable Development Strategy

Vitor Francisco Ferreira*

President, Brazilian Chemical Society Universidade Federal Fluminense,
Instituto de Química, Departamento de Química Orgânica, CEG,
Campus do Valonguinho, Brazil
*E-mail: cegvito@vm.uff.br

In recent years, the changing landscape of science, technology and innovation in emerging nations, such as Brazil, China and India, has had a major impact on the international community, placing these countries within the most promising economies in the globalized world. In this scenario, the Brazilian Chemical Society (SBQ) is committed to contribute to the development of chemistry in Brazil, a country with fantastic natural resources and the potential to become a global leader in sustainable chemistry. The vision for 2025 includes the alignment of SBQ's global strategies with the principles of sustainable development, highlighting the research on biodiversity, alternative sources of energy, and green chemistry. SBQ is committed to playing an important role in the solution of social and environmental global problems, particularly those that afflict public education at all levels.

Introduction

The "Sociedade Brasileira de Química" (SBQ), was founded in 1977 and will celebrate its 40th anniversary in 2017. It is the leading chemical society in Brazil and is devoted to the development and growth of the Brazilian chemical community, the dissemination of chemistry information, and the applications of chemistry to the development of the country and to the improvement of quality of life.

SBQ has discussed with the Brazilian government and other stakeholders the importance of chemistry to the economy of Brazil, always emphasizing that *the chemistry of the future must be clean and sustainable*. This is the vision of SBQ. We believe that innovation, education, science and technology are strategic drivers in ensuring the social and economic advancement of the country.

With respect to clean and sustainable chemistry, SBQ has adopted several global strategies:

1. Research in biodiversity and conversion of biomass in feedstock.
2. Alternative sources for clean energy and green chemistry.
3. Sustainable production of chemicals and drugs.
4. Actions to alleviate social and environmental global problems, mainly those that afflict public education at all levels.
5. Agricultural productivity to feed the increased population.
6. Improvement of the quality of drinking water.
7. Discovery of new drugs for old and new diseases.

These strategies address many of challenges in the world. We know, for example, that by 2050 the world's population will likely reach 9 billion. Currently at least 40 million people in the world have no access to drinking water. Chemistry can help with all of these challenges. In this article, the author will discuss the first two strategies in detail.

Biodiversity and Conversion of Biomass in Feedstock

Brazil is a country with huge natural resources. It has the potential to become a global leader in sustainable chemistry. Nevertheless, it needs to place human beings as the focus for sustainable development, bringing to the discussion issues such as hunger, food production, and diseases.

The chemical industry is important not only in generating jobs but also in contributing to human welfare. In fact, there is no country with sizable economic production that has not also a large chemical production. Furthermore, the countries that are today the most economically dynamic can be easily identified by their prominence in chemical production.

In Brazil chemical industry is in transition. In the past the chemical feedstocks came primarily from fossil materials (e.g., coal, oil, and natural gas). We have learnt in the past 100 years how to convert fossil feedstocks into organic, polymeric, and fine chemicals for commercial use. In the future it is expected that the fossil materials will decrease in importance, and there will be a progressive increase in the use of renewable feedstocks for the chemical industry. We need to learn then how to transform biomass materials into new organic, polymeric and fine chemicals.

Most countries in the world have biomass that can be gathered and utilized. The availability of biomass is a connecting point for all these countries. Perhaps the cultivation or the processing methods are different. Perhaps the plants or the species are different. Yet, most of the biomass contain carbohydrates (including cellulose, hemicellulose, starch, and sucrose), lipids, lignin, proteins, amino acids, chitin, or terpenes. The chemistry of these materials is very rich and may form the cornerstone for the development of new businesses that are important in the future. Because most of these materials are present in all countries, the technologies developed in one country can be applied to another country, thereby permitting fruitful collaborations. Another corollary is that the chemistry professionals need to adapt to this new reality in their work. Educators also need to revise their curricula to give greater emphasis to the chemistry of biomass and its various components.

An example of sucrose may be cited here. Sucrose is the most abundant disaccharide in the world and a major product of Brazil. In Brazil in the 2011/2012 season, 6.9 million tons were produced, mostly from sugar cane. In the same period, 5.12 million tons were exported. Yet, sucrose is also an excellent starting material for a large number of chemical products (Figure 1). The most well known product is bioethanol (to be discussed in the next section). Other chemicals can be useful synthons for the syntheses of drug molecules, specialty chemicals and polymers. There are a lot of opportunities in this area for the creative scientists to come up with good ideas for new products.

Clean Energy and Green Chemistry

Brazil is very active in deploying alternative sources for clean energy and green chemistry. Most of the sources come from either ethanol or lipids. Brazil is outstanding as the world's most intensive user of ethanol from sugarcane as the alternative to gasoline. It is producing 27 billion liters of ethanol from 405 manufacturing plants. The ready availability and low cost of sugarcane makes the bioethanol economically feasible.

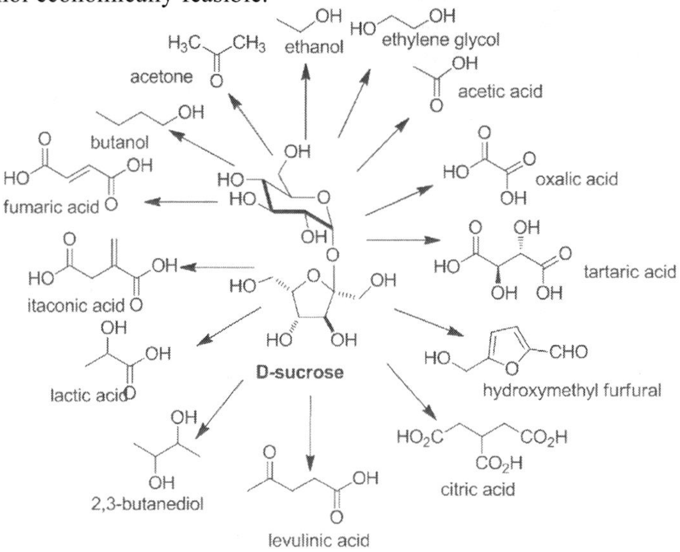

Figure 1. Scheme of reaction pathways from which new products can be made from sucrose.

Another large program in Brazil is biodiesel. Increasingly cars and buses run on biodiesel in Brazil. The main lipid sources for biodiesel production in Brazil are shown in Figure 2. Currently most of it comes from soybean oil.

Main lipid sources for biodiesel production

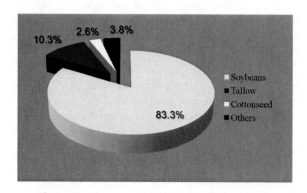

Figure 2. Main lipid sources for biodiesel production in Brazil.

A major consideration when we formulated our strategies is the competition in the production of biofuels versus food on arable land. This is the rationale to convert waste biomass to biofuels. Some examples of the waste biomass being considered are shown in Figure 3.

Generating biofuels from waste biomass eliminates the competition for arable land between biofuels and food sources.

Examples of waste materials:
- Wood from natural forest
- Agricultural residues (straw, stover, cane trash, etc.)
- Agro-industrial waste (sugarcane bagasse and rice husk)
- Animal wastes
- Industrial wastes (paper manufacturing, coconut fiber, orange industries)
- Sewage and city trash
- Food processing wastes (ground coffee, lipids, shrimp shell, etc.)

Figure 3. Examples of waste materials for use in biomass conversion.

Conclusions

From the examples shown above, it is clear that SBQ's vision is to align its global strategies with the principles of sustainable development. Certainly clean energy, green chemistry, and biomass conversion and utilization are among the main themes.

For many years SBQ and ACS have a very good relationship. We hope to continue our interactions and collaborate in areas of mutual interest.

SBQ has also been active in publishing journals and in organizing meetings. In fact, SBQ is hoping to bring IUPAC 2017 to São Paulo. For more information, interested readers may check the following website: http://www.sbq.org.br/IUPAC2017.

Chapter 12

Chemical Sciences: Contributions to Building a Sustainable Society and Sharing of International Responsibilities

Zhigang Shuai,*,[1] Xi Zhang,[1] and Jiushu Shao[2]

[1]The Chinese Chemical Society and Department of Chemistry,
Tsinghua University, 100084 Beijing, China
[2]School of Chemistry, Beijing Normal University, 100875 Beijing, China
*E-mail: zgshuai@tsinghua.edu.cn

In this report, we first demonstrate that chemistry is the central science from the science of matter to the science of life; i.e., chemistry is essential for food supply and resources, for understanding life processes, for providing clean and renewable energy, for material science, for information technology, and for environmental sciences. Then, we summarize the recent advances in chemical sciences in China, in nano science and technology, material chemistry, supramolecular self-assembly, organic synthesis, and more fundamentally physical and theoretical chemistry. Chinese chemists have not only made significant progress in chemical research but also assumed international responsibility in tackling globally challenging problems by collaborating with international chemists and providing chemical solutions to energy, environment, health, and resources problems. Lastly, we present some remaining challenges in chemical sciences and our perspectives on the contributions that chemistry can make towards a sustainable and harmonious society.

Introduction

Chemistry is a science that studies all types of substances, including atoms, molecules, supramolecules, and other large molecular and solid systems. The focus in the field tends to be on syntheses, transformations, isolation, analyses, structural and conformational determinations, and physical and chemical properties of substances. As the most important field that can create new substances, chemistry not only is a central part of fundamental sciences, but also has a broad spectrum of applications. On one hand, chemistry helps us understand why molecules in nature are what they are and tells us how to construct novel substances unavailable in nature. On the other hand, chemistry is essential to material civilization and dramatically improves the quality of life for human beings by providing new materials for numerous applications. Because chemistry is at the center of natural sciences, its advances directly influence biology, medicine, agriculture, materials, and other related areas. There is no doubt that chemistry will continue to play a unique role in advancing developing countries and in building a sustainable and harmonious world.

The development of chemical sciences in all their facets is astonishing. Tens of millions of molecules and compounds have been synthesized and their structures at atomic scale have been elucidated. Novel techniques and synthetic methodologies have been developed, which pave the way to the manufacturing of new materials with targeted properties. In addition to offering methods for the creation of new substances, chemistry also reveals how and why transformations of matter or chemical reactions take place. In fact, the structures and dynamics of chemical systems are favorite research topics in modern chemistry. In view of the principle that structures determine properties, chemical knowledge helps establish structure-property relationships of substances at the molecular, supramolecular, aggregate levels, and beyond, thus achieving the ultimate goal of rational design of new molecules, new drugs and other new substances.

Chemistry has changed the world and benefitted human beings in many ways. A good example is the high-pressure catalyzed process for ammonia synthesis, which allowed large-scale man-made fertilizers to be made. It was one of the most important inventions in the twentieth century, and has revolutionized agricultural production. Another example is oil refinery. The application of highly efficient platinum catalysis has produced vast amounts of high-quality gasoline and aromatic hydrocarbons that are needed by industry. Thanks to chemists' efforts, various analytical instruments are available and are widely used to monitor the environment, and pollution can be effectively traced and controlled accordingly.

The origin of life is one of the most challenging problems in science. Biological processes are complicated due to highly ordered combinations of chemical reactions. The clues to to this complex problem can only be found by studying the chemistry of living things.

Indeed, chemists are getting more and more interested in complex systems at multiple scales. They establish state-of-the-art experimental and theoretical methodologies to explore the structures, dynamics, and functions of the systems. In addition, computers have been extensively employed in modern chemistry, for information storage and analysis, data mining, and theoretical modeling of chemical phenomena. A successful example is the computer-assisted drug design, which tremendously saves money and time when compared with the traditional trial-and-error approach.

Modern chemical sciences have advanced to an unprecedented level, and many tools and methodologies are now available. Certainly chemistry can be used to create new substances to meet the changing needs of human beings. Moreover, the development of new chemical ideas and methods has helped life sciences, material science, and even information technology. Today we are facing global challenges in exploring energy and other resources, in protecting environment, in coping with global warming, in addressing health and medical issues, in maintaining a sustainable world, and many others. The following pages outline the different roles that chemistry can play in addressing these global challenges.

1. Chemistry is one of the key sciences for solving the world's food supply and resources problem.

China is an agriculture-intensive country. A strong agriculture is necessary for a healthy national economy and a guarantee of a stable government. The total population in China will reach 1.6 billion by the middle of this century. It is a formidable task to provide sufficient food supply for this large population and at the same time to protect the agro-ecological environments. Chemistry contributes to the manufacturing of highly efficient and yet environmental friendly fertilizers, pesticides, and plastic membranes for agricultural production. In particular, chemistry is used to design and produce green bio-fertilizers, bio-pesticides, and biodegradable materials. Chinese chemists have made enormous efforts in maintaining and treating soil pollution, desertification, drought, as well as salinization and other cultivation and ecological problems. Both Chinese chemist and scientist from other disciplines will explore the mechanism of the photosynthetic reactions of plants, which will be invaluable as we try to increase the agricultural production via genetic and genomic engineering.

2. Chemical process is fundamental for understanding life so that chemistry can ensure the living standard and health of humankind.

Chemists and the chemical sciences have been integral to the development of modern medicine, from drug compounds to drug delivery systems to diagnostic technologies. The result has been a steady improvement in our health and life expectancy over the past 100 years. Complex chemical processes are involved in the life processes including growth, reproduction, aging, disease, death and so on. The research activities therein are dependent on chemical theory, techniques and methodologies at the molecular level. Chemistry of life, including chemistry of brain and memory is a big challenge and of utmost importance. One can never overestimate the importance of chemistry in medicine. Chemists invent new medicines such as antiviral and antibiotics to relieve pain caused by sickness and to cure diseases. They improve new diagnostic methods so that diseases can be detected, identified, and treated at an early stage. They are also developing biocompatible materials for organ transplant.

103

3. Chemistry is essential for providing clean and renewable energy for the future.

Energy is essential to us. We need energy for manufacturing, transportation, heating, lighting, and daily life. In China, more than 90% of energy utilization rely on chemical processes. It is a great challenge for chemists to design effective chemical processes to transform the low-quality fuels to energy at low costs and reduced pollution. Fossil fuels, including coal, petroleum, and natural gas, are limited in resource and cannot be regenerated. With current technology, the burning of fossil fuel produces a large amount of oxides of sulphur and nitrogen, and other pollutants. Therefore, while efforts in more efficient and cleaner conversion of fossil fuels should be sought, new energy sources should be researched. Chemists need to develop new tools, concepts, and technologies to explore solar energy, nuclear energy, hydrogen and fuel cells, and others.

4. Chemistry consists of the source of materials sciences.

Materials are needed for a wide variety of applications. The national economy, industrial modernization, and national defense all benefit from the developments of new or improved materials. Chemists make new materials by synthesizing new compounds and characterizing their structures and properties. At the atomic, molecular, and supramolecular levels, chemical sciences help to design materials with required optical, electrical, magnetic, and mechanical properties. By virtue of the structure-property relationships, molecular scientists can apply the rational design strategy to develop new generations of semiconductors, optical materials, magnetic materials, superconductors, high temperature heat resistant matters, super-hard materials, and other advanced materials. From synthesis to processing to commodity manufacturing of materials, the tools of chemical science and engineering will be essential. New materials with predictable properties will provide formidable targets for design and synthesis, while processing and manufacturing of these new materials will present new challenges and opportunities for chemical engineering.

5. Chemistry in Information Technology

Information technology is a strong pillar for the 21st century industry. The fast growth and widespread application of internet communication technology has a tremendous impact on human life. Chemistry not only provides basic substances for information generation, transmission, storage, and display, but also provides new materials and manufacturing processes for ultra-large scale integration circuits. Miniaturization of device will ultimately reach molecular size. Electronic devices at molecular level and biochips are among the most rapidly growing areas with an intense focus in chemistry. Among them, molecular wires, molecular switches, molecular motors, and molecular rectifiers have achieved great success recently. The development of scanning probe microscopy (SPM) allows chemists to study single atom, single molecular behavior, as well as electronic device performance at molecular scale.

6. Chemistry can provide tools and methods to address environmental problems.

As the population increases, urbanization spreads extensively, energy consumption soars, and pollution in water, atmosphere, and soil becomes very serious. Chemists need to find economically viable reactions and design environmentally benign reaction processes that can be scaled up from laboratory to industry. Green chemistry is the ultimate solution. In addition, chemists need to understand the chemical compositions and physical phenomena of the earth. They need to understaned the chemistry that occurs in rivers, lakes, oceans, and atmosphere as well. With improved knowledge, chemists can tackle environmental problems. For instance, they can 1) establish highly sensitive detection methods to monitor the migration and transformation of pollutants. 2) Develop efficient and clean approaches for fuel burning. 3) Design suitable processes to remedy polluted water and soil. Of course, chemists should be responsible for instituting and revising the standards for environment protection management.

In summary, chemical sciences not only made great contributions in the past, but also await a more exciting future. As chemists, we should be ready to meet the ever-increasing challenges in managing the world's resources, energy, water, climate, and environment.

Recent Advances in Chemical Research in China

The year 1978 was called "Springtime for Science in China" and a very special year in the scientific history of China: The authorities of the Central Government started to recognize that science and technology are the main driving forces for China's modernization and inaugurated an unprecedented National Science Conference. Since then, the scientific endeavors in China have progressed at a rapid pace. The warmth of the Springtime hopefully will continue to be beneficial to the scientific research and development in China, and at the same time it will bring big rewards. The growth of chemical sciences in China is reflected by the increase in the number of publications. According to the statistics from Chemical Abstract (CAplus), China has always ranked No. 1 in terms of total number of publications in chemistry for the period of 2006-2010 shown in Table 1.

In fact, since 1996 many Chinese scientists have returned home with research experiences, either in Ph.D. programs and/or as postdocs in US, Europe, and Japan. To attract more young and experienced scientists from abroad, the Chinese Academy of Sciences and some elite Chinese universities have launched various "Talent Search Program" since then. The most successful example is the so-called "Hundred Talents Program" supported by the Chinese Academy of Sciences. Since 1995, about 2500 scientists from US, Europe, Japan, and other countries have joined its research institutes and many of them have become leading figures in their fields. Besides, the National Natural Science Foundation of China (NSFC), the most important funding agency for basic science in China, started the "Outstanding Young Scientist Award" (OYSA) program in 1994, aiming at supporting potential Chinese scientists under age 45 to start up their ambitious research groups. The OYSA program has been extremely successful and has been considered as a great honor for all young Chinese scientists. In particular, each awardee receives a considerable amount of funding for a period of 4 years. About 500 young chemists have won OYSA awards until now and they have become the research leaders responsible for developing chemical sciences in China.

Table 1. Top 10 countries in terms of total publications in Chemistry for 2006-2010[1]

Country	2006-2010	Rank	Proportion	Growth rate
P. R. China	117420	1	18.59%	14.35%
USA	114514	2	18.13%	2.13%
Japan	52978	3	8.39%	-1.19%
Germany	48265	4	7.64%	1.56%
India	38920	5	6.16%	8.21%
France	33309	6	5.27%	1.26%
England	31126	7	4.93%	0.11%
Russia	28553	8	4.52%	-1.10%
Spain	24983	9	3.96%	3.23%
S. Korea	21933	10	3.47%	7.47%

[1] From CAplus

Undoubtedly, it is the people making all scientific achievements possible. It is also true that the performance and breadth of talented scientists, especially experimentalists, can be greatly enhanced with strong support. The story is the same in China: scientific research and scientists benefit a lot from the booming economy. The funding for basic science has been increased more than 10 folds during the past 15 years. Tremendous improvements have been achieved in terms of laboratory facilities, space and conditions, and the workforce. As noted by Professor Peter Stang of the University of Utah, Editor of the Journal of the American Chemical Society: that "Chinese chemists are already world class in some areas." Indeed, some top-level chemistry research institutions such as the Institute of Chemistry of the Chinese Academy of Sciences and the College of Chemistry and Molecular Engineering of Peking University aim at becoming world first-class research centers for chemical sciences within 10 years. To this end, these two institutions decided to unite their forces to form the Beijing National Laboratory for Molecular Science (BNLMS), which has obtained strong support from the Ministry of Science and Technology of China with an annual budget of about 6 millions USD.

J. Am. Chem. Soc. and Angew. Chem. Int. Ed. are widely regarded as the top chemistry journals. The following charts (Figure 1 and Figure 2) show the impressive increases for the papers in these two journals published by Chinese chemists in the period of 2000 - 2012.

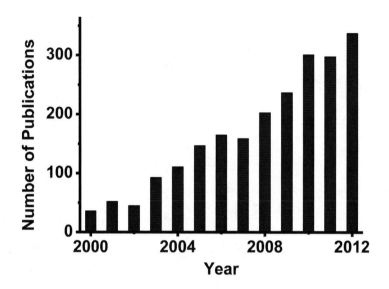

Figure 1. Yearly publications in JACS by Chinese chemists. (from Essential Science Indicators Web of Science.)

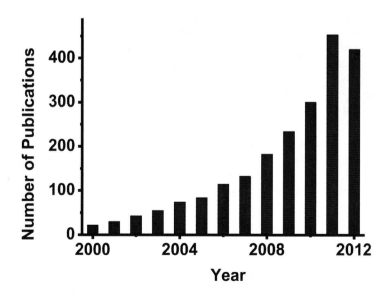

Figure 2. Yearly number of publication in ACIE by Chinese chemists. (from Essential Science Indicators Web of Science.)

It is evident that not only the quantity, but also the quality greatly improved in the past decades. This is also manifested by the total number of citations (see Table 2).

Table 2. Total number of citations for chemistry papers in the period of 2006-2010. (from Esential Science Indicators Web of Science)

Country	2010	2006-2010	Rank
USA	20872	990901	1
P. R. China	10085	466396	2
Germany	7682	345650	3
Japan	5177	300426	4
England	5372	233100	5
France	4146	212280	6
Spain	3256	166266	7
India	3083	142638	8
Italy	2714	134302	9
Canada	2316	119582	10

It is not the purpose of this report to make an overall assessment on the chemical research progress from China. We intend to present some representative research results in China in nano-science, molecular and supramolecular self-assembly, chemistry of advanced materials, environmental chemistry, organic synthesis, and fundamental physical and theoretical chemistry.

1. Advances in Nano-Science

An important milestone of nanoscience in China is the development of home-made scanning tunneling microscope (STM) by Prof. Chunli Bai from the Institute of Chemistry of the Chinese Academy of Sciences. In the mid-1980s the STM was not yet commercially available. He successfully designed and developed not only the very first STM, but also the first atomic force microscope (AFM), low-temperature STM, ultra-high vacuum-STM and the ballistic electron emission microscopy (BEEM) in China. Indeed, these were the earliest technological tools in the country for characterizing and manipulating single atoms and molecules as well as surfaces and interfaces in the nano-scale world. These achievements marked the birth of scanning probe microscope (SPM) research and opened the door to nanoscience research in China. Owing to Bai's efforts, nanoscience has been a fast growing field in the last twenty years.

Print industry is one of the backbone industries in China; however, it produces a large amount of solid waste, air emission, and wastewater. In order to avoid the complex multi-step processing of laser typesetting plate-making and serious pollution of traditional printing industry based on photosensitizing process, Prof. Yanlin Song from the Institute of Chemistry of the Chinese Academy of Sciences invented a non-photosensitizing, non-pollution and low-cost green plate-making technology (*1*, *2*). Based on the manufacturing of functional nanomaterials, controllable spreading and transferring of liquid droplets, they fabricated the superoleophilic patterns on the hierarchically structured superhydrophilic plate by ink-jet printing (*3*, *4*). Since, the the photosensitizing process is completely avoided, the new technology simplifies the chemical engineering process, discharges no chemical pollluants, and reduces cost (Scheme 1). In a step forward, they employed such green printing technology to printing electronics based on metal nanoparticle ink, which simplifies the traditional photolithography method, and reduce discharge of chemical pollutant (*5*). The necessary technologies for the green plate-making and printing as-a-whole have been pooled together. Pilot lines have been established, which has an annual production capability of millions of square meters of printing plates and tens of metric tons of nanomaterials.

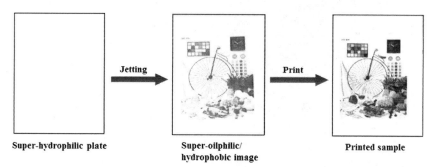

Super-hydrophilic plate   Super-oilphilic/ hydrophobic image   Printed sample

Scheme 1. Schematic demonstration of the production process, which shows that no photosensitizing process is involved in this new technology and the printing plate is directly obtained by inkjeting functional nanomaterials on the superhydrophilic plate to achieve desired superhydrophobic image area.

The ultimate goal of chemistry is to shape the world at the single molecular level. Among all properties of matters, the magnetism of a single molecule is particularly interesting because of its potential applications in future molecular scale information storage and quantum devices. Profs. Jianguo Hou, Jinlong Yang and coworkers from the University of Science and Technology of China reported the first example of manipulating the magnetism of a single molecule with a scanning tunneling microscope, an extraordinary example of bond-selective chemistry (*6*). It is known that there is an unoccupied spin in a free cobalt phthalocyanine molecule, but the magnetism is completely quenched when the molecule is adsorbed on an Au (111) surface. Cutting away eight hydrogen atoms from the molecule with voltage pulses from a scanning tunneling microscope tip allowed the four lobes of this molecule to chemically bond to the gold substrate, and the localized spin was then recovered in this artificial molecular structure (see Figure 3). This beautiful experiment was highlighted by the Science magazine, which reported "*the result opens the way for fundamental studies of spin behavior in molecules that may influence future molecular device applications*". The work clearly *demonstrates an ability to change the magnetic state of a molecule by directly modifying its structure via single-molecule manipulation*; it *takes such manipulations to a new level.*

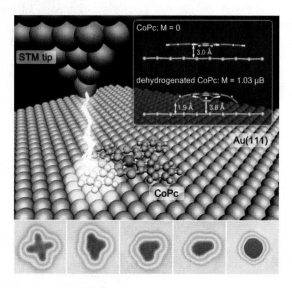

Figure 3. Illustration of manipulating single molecule magnetism. (see color insert)

Because of the complexity of structures and properties of different materials, it is extremely difficult if not impossible to develop a universal methodology for the controlled growth of low-dimensional nanostructures. Prof. Yadong Li of Tsinghua University has invented an all-around method, namely the liquid-solid-solution phase transfer and separation process, for creating nanocrystals (7). Their synthetic scheme was based on a general phase transfer and separation mechanism occurring at the interfaces of the liquid, solid and solution phases present during the synthesis (Figure 4). This method can be feasibly used to produce nanocrystals of noble metals, semiconductors and conducting polymers, that is, magnetic, dielectric, fluorescent, optoelectronic or biomedical nanocrystals. This novel synthetic procedure enables much progress in understanding the intrinsic size-dependent properties in different systems of nanocrystal building blocks and drives more unique and exciting applications in the bottom-up nanotechnology.

Prof. Lijun Wan from the Institute of Chemistry of the Chinese Academy of Sciences has made remarkable contributions in controlling the distribution and dispersion of organic/biomolecules on solid surfaces, which is indispensable for nanomaterials and nanotechnology. Using self-assembled technique, he has successfully fabricated a molecular template of end-functionalized oligo(phenylene-ethynylene) (OPE). The structure and molecular arrangements in the template were clearly determined by the electrochemical STM. With the molecular template, organic molecules such as coronene (COR) and biomolecules such as tripeptides are well distributed and are monodispersed on highly oriented pyrolytic graphite (HOPG) surfaces. COR molecules were controllably distributed into the template and self-organized into various arrays by simply adjusting the molecular molar ratio. Recently, they reported the induction of global homochirality in two-dimensional enantiomorphous networks of achiral molecules via co-assembly with chiral co-absorbers. The STM investigations and molecular mechanics simulations demonstrate that the point chirality of the co-absorbers transfers to organizational chirality of the assembly units via enantioselective supramolecular interactions and is then hierarchically amplified to the global homochirality of two-dimensional networks (8).

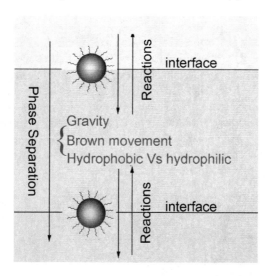

Figure 4. A novel general synthesis method for nanocrystals. (see color insert)

While stable fullerenes, such as C_{60} and its larger homologs, have been macroscopically synthesized since 1990, the fullerenes smaller than C_{60} are highly reactive and have eluded us for twenty years. As an important achievement in the smaller fullerene synthesis, a group of researchers led by Lan-Sun Zheng at Xiamen University in China stabilized an elusive C_{50} cage in milligram amounts by chlorination (9). The structure of $C_{50}Cl_{10}$ was characterized by mass spectrometry and nuclear magnetic resonance (NMR), and confirmed by X-ray crystallography recently (Figure 5). This research group initiated their studies on carbon clusters in gas phase since 1980s, and entered the scientific field of macroscopic synthesis of fullerene-related compounds in 1990s. Utilizing various plasma-generating technologies combined with chlorination reactions, they have now been able to synthesize a series of chlorinated derivatives of other labile fullerenes, e.g., C_{54}, C_{56}, C_{58} and isomeric C_{60}'s. These works provide the practical route to the bulk synthesis of the elusive fullerenes and their derivatives, and provide new insights into the mechanism of fullerene formation.

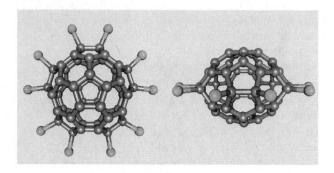

Figure 5. Schematic structure of $C_{50}Cl_{10}$. (see color insert)

Prof. Shi Gang Sun of Xiamen University in collaboration with Prof. Zhong Lin Wang of the Georgia Institute of Technology, USA, has used electrochemical synthesis to create a highly efficient new class of multifaceted catalysts, *i.e.* platinum nanocrystal catalysts with 24 facets, a breakthrough in the synthesis of nanoscale catalysts (*10*). These tetrahexahedral nanoparticles have high-index facets with unsaturated surface areas that help make the catalysts up to 4.3 times more efficient than spherical platinum nanoparticles (per unit platinum surface area) at oxidizing organic fuels such as formic acid and ethanol (Figure 6). In addition, the nanoparticles are remarkably robust and can remain stable at temperatures up to 800 °C, which makes them recyclable in relevant applications.

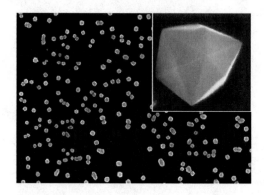

Figure 6. STM image of tetrahexahedral Pt nanocrystals (10).

2. Progresses in Advanced Materials Research

The theoretical research on the structural chemistry at Fujian Institute of Research on Structure of Matter, Chinese Academy of Sciences provided applicable guidelines to the discovery of new functional materials for optical applications, which posed a successful story of rational design of nonlinear optical (NLO) crystals. Theoretical studies on the relationship between crystal structure and the NLO properties have directly led the way for the discovery of β-BaB$_2$O$_4$ (BBO) and LiB$_3$O$_5$ (LBO) (Figure 7). These achievements have paved the way for frequency conversion in both visible and UV regions. Large sized crystals with high optical quality are now available, thanks to the improvement of crystal growth technology. BBO and LBO crystals have become the major operating materials in today's solid state laser systems, and the mainstream in commercial opto-electronic products in both civil and military applications. The example shows a very successful case of transforming highly original research in fundamental chemistry to wide applications.

Figure 7. Large-sized nonlinear optical LBO crystals.

Organic solids are pi-conjugated functional molecules and polymers, charge transfer salts, and their nanostructures. They possess great potential in applications covering but not limited to display and lighting, field-effect transistors and printable electronics, solar energy conversions, bio- and chemical-sensing, information switching and storages, and anti-corrosion coatings, to name a few. Since 1980's Prof. Daoben Zhu from the Institute of Chemistry of the Chinese Academy of Sciences has successfully established a center of excellence in this field, namely, the Key Laboratory of Organic Solids. This laboratory, consisting of 14 independent research groups now, has gradually become a global elite research center in organic solids. Each year, more than 30 publications in the top chemistry journals such as *J. Am. Chem. Soc., Angew. Chem. Int. Ed., Adv. Mater* are produced from this laboratory. Zhu's laboratory is among the best in the world in organic semiconductor synthesis and device fabrication in the field of organic thin film transistors, organic light emitting diodes, organic solar cells, chemical- and bio-sensors, organic photonics crystals, as well as in bio-inspired interfacial materials. Recent advances include: (i) preparation of a number of novel organic semiconductors with mobility ~ 10 cm²/Vs, including first demonstration of conjugated polymers with unprecedented high mobility of 8.2 cm²/Vs and donor-acceptor alternatively co-crystalization which enable ambipolar and air-stable transports (*11, 12*). The miniaturization has been realized with nanocrystal organic devices (Figure 8) (*13*); (ii) Light-emitting polymers containing either cations or anions in the side chains were employed as biosensors to detect the conformational change of DNA, enzyme activity, and more biomolecules (*14, 15*). (iii) Novel n-type polymers and fullerene derivatives (ICBA) have been synthesized for better performance of organic solar cells (*16*).

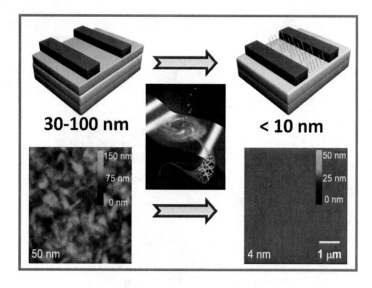

Figure 8. Organic electronic device can be made ultrathin at nanoscale. (see color insert)

Prof. Yong Cao group in South China University of Technology simplified polymer light-emitting diodes fabrication process by printing Ag-conducting paste on cationic conjugated polyelectrolyte surfcae as a bilayer cathode. This printing technology makes it possible to fabricate polymer light emitting device and displays without the use of thermal deposition, thus creating an avenue to achieve all printable roll-to-roll polymer light emitting devices and displays. When amino alkyl functionalized polyfluorene polyelectrolyte (PFN) is layered on RGB electroluminescent (EL) polymers, mixing does not occur since EL polymers are not soluble in methanol. Cao's group found that cationic water-/alcohol-soluble polyelecrolytes and their precursors allow efficient electron injection from high work function metals such as Al , Ag, and Au. Ag-paste cathode can also be printed on the top of PFN layer and the resultant PLEDs showed (device efficiency and EL spectra) comparable device performance to those with Ba, Ca and PFN/Ag (by thermal deposition) cathodes. The best luminous efficiency reached 7.8 and 5.6 cd/A for the green-emitting P-PPV and blue-emitting PFO devices with Ag-paste cathode, respectively. This is the first reported polymer light-emitting devices fabricated exclusively by printing technology without thermal deposition involved and a big step forward to make all printable roll-to-roll polymer light-emitting devices, displays and illumination sources (*17, 18*). Their recent advance includes the invention of inverted structure for polymer solar cells; with their D-A copolymer blended with PCBM, power conversion efficiency reached a record high of 9.2% (*19*).

3. Advances in Molecular and Supramolecular Self-Assembly

Besides the wide panel of physical properties provided by functional compounds, organic nanostructures also exhibit a wide range of optical and electronic properties that depend sensitively on both their shapes and sizes. Prof. Jiannian Yao and coworkers in the Institute of Chemistry of the Chinese Academy of Sciences demonstrated the excellent features of organic nanostructures in lasing, waveguiding, multiple emissions and color tenability. Once integrated into a functional device, organic nanostructures should have a bright future. Although the tailor-made molecules can generally be obtained by organic synthesis, generating molecular aggregates with a specific structure and nanostructures with a desirable function remains a great challenge. Yao, et al. focused on the rational fabrication of organic nanostructures with a controllable shape, size, and therefore function. Aiming at a general strategy for creating well-defined supramolecular objects, they revealed how the molecular structures of building blocks affect the morphology of nanodimensional assemblies, and developed a kinetic control method for preparing 0D and 1D organic nanoheterojunction structures (Figure 9). Constructing highly-ordered superstructures through self-organization of organic nanostructures opens new routes to organic optoelectronic devices in a cost effective way (*20, 21*).

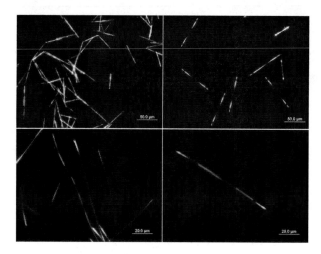

Figure 9. Yao et al. prepared organic nanoheterojucntions for optoelectronics application through molecular self-assembly. (see color insert)

Prof. Dongyuan Zhao of Fudan University in Shanghai has, *via* EISA strategy, created a series of highly ordered mesoporous polymers and carbons with 2-D hexagonal and 3-D cubic structures (*22*). The precursor is phenolic resol and the template is a mixed amphiphilic surfactant system of PEO-PPO-PEO and PPO-PEO-PPO. The mixed block copolymers interact with resols and assemble into cross-linked micelles that are suitable templates for constructing mesostructures. The interface curvature of the cross-linked micelle depends on the chemical composition, hydrophilicity/hydrophobicity as well as the specific feature of the reverse PPO-PEO-PPO (Figure 10).

115

Hyperbranched polymers (HBPs) are a new type of macromolecules with spherical and highly branched topology and a large population of terminal functional groups. Prof. Deyue Yan and co-workers of Shanghai Jiaotong University have prepared many types of HBPs self-assemblies at all scales and dimensions (Figure 12) (*23, 24*). These self-assemblies can be macroscopic tubes, physical gels, micro- or nano-vesicles, fibers, spherical micelles, honeycomb films and large compound vesicles, depending on the media in which they formed. In addition, Yan's group have found that the vesicles self-assembled from HBPs possess good membrane fluidity like liposomes and strong stability like polymersomes, serving as a simple model membrane to mimic cellular morphologies and functions. Indeed, they have observed membrane fusion, fission, swelling, and shrinkage by using the polymer vesicles self-assembled from a hyperbranched polyether (Figure 11). These findings help lead to the discovery of new supramolecular structures as well as the new applications, and shed lights on the underlying principle of self-assembly in nature.

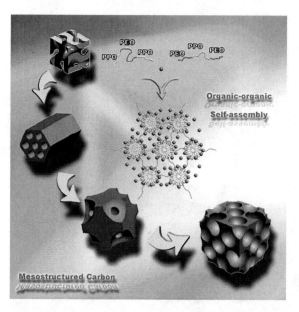

Figure 10. Mesostructured carbon. (see color insert)

The development of new building blocks and creation of diversified supramolecular nanostructures have always been the most important parts in the area of self-assembly. In contrast to conventional amphiphiles based on covalent bonds, Prof. Xi Zhang of Tsinghua University developed a new field of supra-amphiphiles that refer to amphiphiles constructed on the basis of noncovalent interactions or dynamic covalent bonds (*25–27*). In supra-amphiphiles, functional moieties can be attached by noncovalent synthesis, greatly reducing the need for tedious chemical synthesis (Figure 12). The building blocks for supra-amphiphiles can be either small molecules or polymers. The advance of supra-amphiphiles will not only enrich the family of conventional amphiphiles but also provide a new kind of building blocks toward complex self-assemblies, including hierarchical self-assemblies and functional nanostructures.

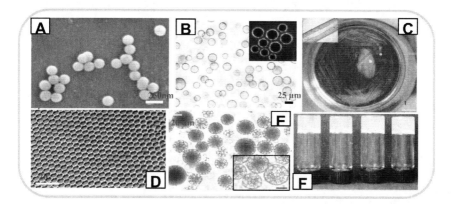

Figure 11. Self-assembly of hyperbranched polymers. (A) spherical micelles; (B) giant vesicles; (C) macroscopic multiwalled tubes; (D) honeycomb-patterned films; (E) large compound vesicles; (F) physical gels. (see color insert)

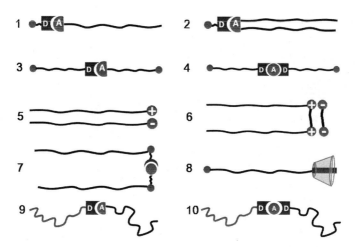

Figure 12. Molecular Engineering of supra-amphiphiles.

4. Advances in Organic Synthesis

The efficient and highly selective synthetic method is always one of the holy grails in chemistry. Profs. Chen-Ho Tung, Lizhu Wu and their co-workers utilized molecular aggregates, cavities and surfaces of microporous solids as microreactors to successfully control the product selectivity in organic photochemical reactions. They devised a new procedure for the preparation of large-ring compounds with high yields under high substrate concentrations by using microreactors (Scheme

2). They could direct the photosensitized oxidation of alkenes selectively towards either the singlet oxygen mediated or the superoxide radical anion mediated products by controlling the status, the location of the substrate and the sensitizer molecules in the microreactor. A number of products that are difficult to prepared in homogeneous solutions are easily synthesized using this mathod (*28–30*).

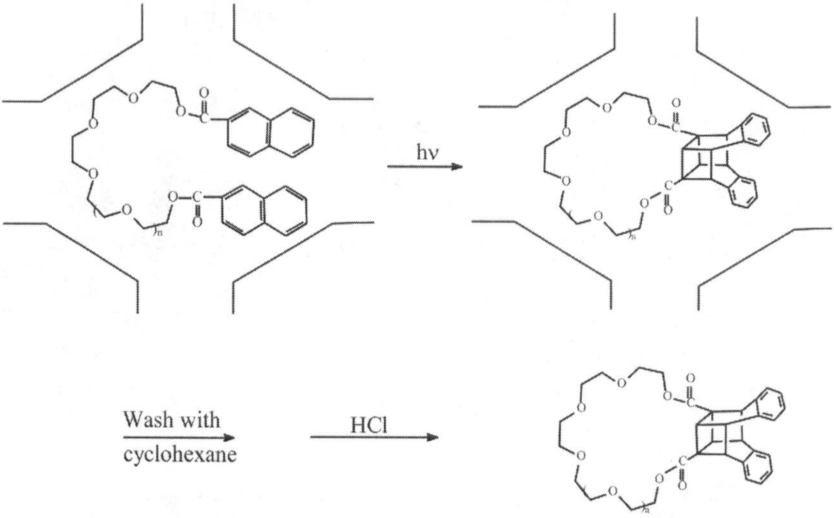

Scheme 2. Micro-reactor for photochemistry synthesis. (see color insert)

Allenes are a class of compounds with interesting reactivity due to their unique structures. They were considered to be "highly unstable" for a long period of time. Prof. Shengming Ma's group has played a major role in allene chemistry, discovering the cyclization of functionalized allenes and the addition reactions of allenes. They found that the two- or three-component reactions can very efficiently produce poly-substituted butenolides, vinylic epoxides, 2,5-dihydrofurans, furans among many others. His group has demonstrated that 1,5-bisallenes can undergo cyclometalation in one step to form complicated skeletons including the steroid-like derivatives with highly stereoselectivity. Nucleophilic conjugate addition reactions of 1,2-allenyl carboxylic acids, esters, amides, and ketones have been developed for the high regio- and stereo-selective synthesis of functionalized alkenes, which are usually difficult to obtain due to the potential migration of the C=C bond. His group has also developed the halo- or seleno-hydroxylation of 1,2-allenyl sulfides, selenides, sulfoxides, sulfones, phosphonates, phosphine oxides, beta-(1,2-allenyl)butenolides, with very high regio- and stereoselectivity. These contributions render allenes very useful in organic synthesis (Scheme 3).

Transition metal-catalyzed asymmetric catalysis is one of the most powerful methods for producing optically active chiral building blocks used in the synthesis of natural products, chiral drugs, agrochemicals, and chiral materials in an environment-friendly and sustainable manner. In the study of transition metal-catalyzed asymmetric catalysis, the design and synthesis of efficient chiral ligands and catalysts is the central goal. Since the beginning of this century, Prof. Qilin Zhou at Nankai University developed a new type of chiral ligands based on the novel 1,1'-spirobiindane backbone including diphosphines SDPs, bisoxazolines SpiroBOXs, phosphine-oxazolines SIPHOXs, a wide range of monodentate phosphorous ligands such as SIPHOS and others (*31–35*). These spirobiindane ligands and the corresponding catalysts possess the advantages of high rigidity and stability, perfect C2-symmetry, as well as easy modification. The chiral spiro ligands and catalysts have been demonstrated to be highly efficient and enantioselective for many asymmetric reactions such as asymmetric hydrogenations of enamines, imines, unsaturated carboxylic acids, and simple ketones, asymmetric insertions of carbenes into heteroatom–hydrogen bonds (X−H; X = O, N, S, Si), and asymmetric carbon-carbon bond-forming reactions, see Scheme 4. These chiral spiro ligands and catalysts have been accepted as a privileged group of chiral ligands and catalysts, and this work is considered an extraordinary contribution to chiral science and technology.

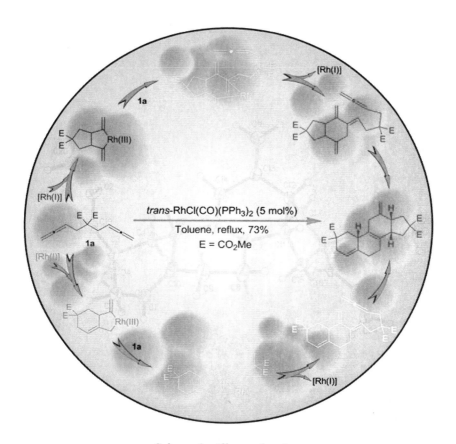

Scheme 3. Allenes chemistry.

119

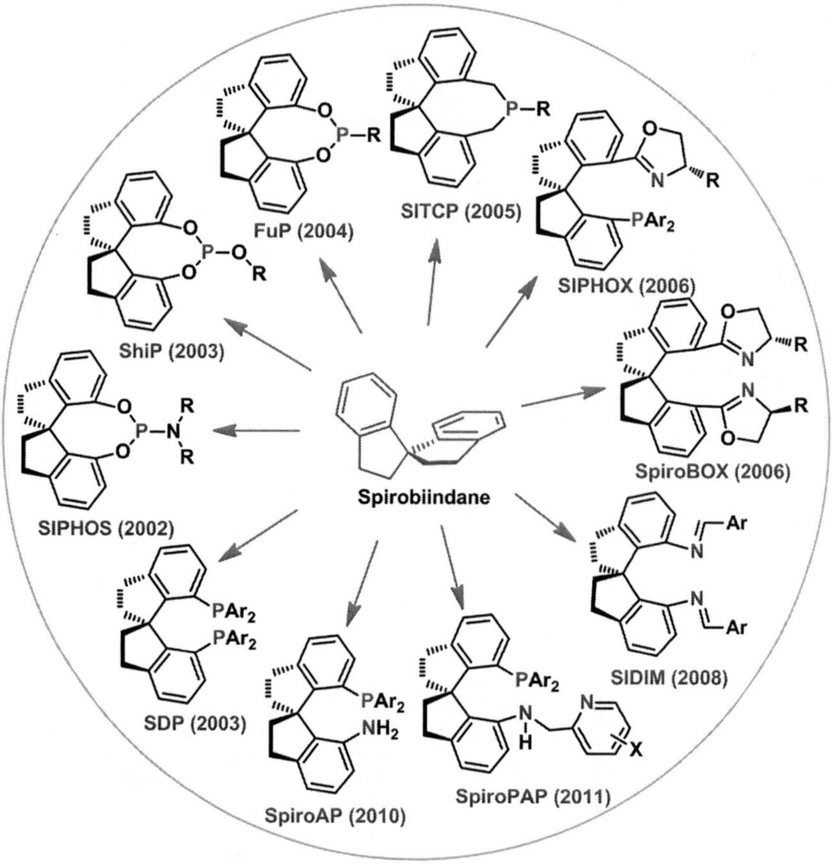

Scheme 4. Example of chiral spiro ligands.

5. Advances in Environmental Chemistry

It has been previously shown that hydroxyl radicals (HO•) can be produced by H_2O_2 and halogenated quinones, independent of transition metal ions. However, the underlying molecular mechanism is still unclear. Prof. Gui-Bin Jiang and coworkers from the Research Center for Eco-Environmental Sciences of the Chinese Academy of Sciences, found that it is tetrachloro-1,4-benzoquinone (TCBQ), instead of its corresponding semiquinone anion radical, the tetrachlorosemiquinone anion radical (TCSQ•-), that is essentially responsible for HO• production. They propose a novel mechanism: a nucleophilic attack of H_2O_2 onto TCBQ, forming a trichloro-hydroperoxyl-1,4-benzoquinone (TrCBQ-OOH) intermediate, which decomposes homolytically to produce HO•, see Scheme 5. It has also been found that the halogenated quinones could enhance the decomposition of organic hydroperoxides and formation of alkoxyl radicals in a similar pathway. These findings may explain the carcinogenicity of the widely used biocides including polychlorinated phenols (such as wood preservatives pentachlorophenol, 2,4,6- and 2,4,5-trichlorophenol), hexachlorobenzene, and

Agent Orange (the mixture of 2,4,5-trichlorophenoxyacetic acid (2,4,5-T) and 2,4-dichlorophenoxyacetic acid (2,4-D)), because these compounds can be metabolized *in vivo* to tetra-, di- or mono-chlorinated quinones (*36, 37*).

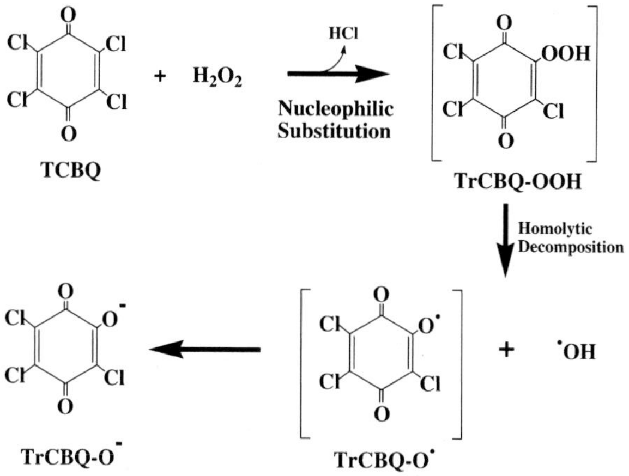

TCBQ + H_2O_2 → (Nucleophilic Substitution, HCl) → TrCBQ-OOH

TrCBQ-OOH → (Homolytic Decomposition) → TrCBQ-O· + ·OH

TrCBQ-O· ← TrCBQ-O·

Scheme 5. Proposed mechanism of HO radical production by TCBQ and H₂O₂ (37).

To promote efficient use of solar energy, many efforts have been devoted to the modification of TiO₂ to extend its spectral response to visible region. A group led by Prof. Jincai Zhao in the Institute of Chemistry of the Chinese Academy of Sciences has successfully improved the photocatalytic activity in the visible region. Through doping the nonmetal element boron and the metal oxide Ni₂O₃ in TiO₂, toxic organic pollutants such as trichlorophenol (TCP), 2,4-dichlorophenol (24-DCP) and sodium benzoate can be efficiently degraded and mineralized. A density functional theory calculation by Prof. Zhigang Shuai showed that boron doping can induce mid-gap band, thus extending the absorption range. The dechlorination and mineralization results point towards the photocatalytic pathway via visible light excitation (*38*).

6. Advances in Fundamental Physical and Theoretical Chemistry

Physical and theoretical chemistry is the foundation of chemical sciences. Resonances in the transition state region are important in many chemical reactions near the reaction energy thresholds. Detecting and characterizing isolated reaction resonances, however, have been a major challenge in both experiment and theory. A research group led by Profs. Xueming Yang and Dong Hui Zhang from Dalian Institute of Chemical Physics of the Chinese Academy of Sciences has carried out a series of high resolution crossed-molecular-beams scattering experiments on the F+H₂/HD reaction with product quantum states fully resolved (*39*), in combination with quantum dynamics calculations on accurate potential energy surfaces. Pronounced forward scattering for the HF(v'=2) product has been observed at the collision energy of 0.52 kcal/mol in the F+H₂(j=0) reaction (Figure 13). Quantum dynamical calculations based on new potential energy surfaces show that the forward scattering of HF(v′=2) is due to two Feshbach resonances. Quantum state resolved scattering on the isotope substituted F+HD→HF+D reaction has been studied. A remarkable, fast changing dynamical picture has been observed, providing an extremely sensitive probe to study the resonance picture of this benchmark system. Furthermore, forward scattering HF(v′=3) product from the F+H_2 reaction was also observed and attributed mainly to a slow-down mechanism over the centrifugal exit barrier. More interestingly, based on the theoretical prediction made on an accurate potential energy surface, the group observed a clear oscillatory structure assigned to J = 12 to 14 partial waves in the collision energy dependence of the state and angle-resolved differential cross sections (Figure 14) (40). From these concerted experimental and theoretical studies, a clear physical picture of the reaction resonances has emerged, providing a textbook example of dynamical resonances in elementary chemical reactions.

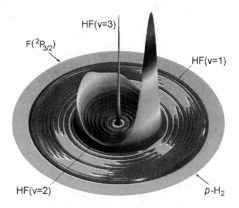

Figure 13. The 3D contour plot of the experimental differential cross sections for the F+H2 reaction at the collision energy of 0.52 kcal/mol.

Surface-enhanced Raman scattering (SERS) is a powerful spectroscopy technique that can provide non-destructive and ultrasensitive characterization down to single molecular level, comparable to single-molecule fluorescence spectroscopy. However, general substrates based on metals such as Ag, Au and Cu, either with roughened surfaces or in the form of nanoparticles, are required to realize a substantial SERS effect, and this has severely limited the breadth of practical applications of SERS. Tian group in Xiamen University report an approach, named shell-isolated nanoparticle-enhanced Raman spectroscopy, in which the Raman signal amplification is provided by gold nanoparticles with an ultrathin silica or alumina shell. A monolayer of such nanoparticles is spread as 'smart dust' over the surface that is to be probed. The ultrathin coating keeps the nanoparticles from agglomerating, separates them from direct contact with the probed material and allows the nanoparticles to conform to different contours of substrates. They have been able to obtain high-quality Raman signal from various species on a broad range of samples, including single crystal metal and semiconductor surfaces, fruits, living cells. SHINERS method has significantly expanded the flexibility of SERS for useful applications in materials and life sciences, as well as for the inspection of food safety, drugs, explosives and environmental pollutants. By placing a sharp Au or Ag tip with high surface plasmonic resonance quality close to a sample surface, the Raman signal of the sample can also be significantly enhanced right in the vicinity of the tip. This method provides high spatial resolution and high sensitivity (Figure 15). They also developed a method called fishing mode tip-enhanced Raman spectroscopy, to significantly increase the chance to form tip-molecule-substrate junction, and to allow mutually verifiable single-molecule conductance and Raman signals with single-molecule contributions to be acquired simultaneously at room temperature. In particular, they revealed that a stronger bonding interaction between the molecule and tip may account for the nonlinear dependence of conductance on bias voltage. FM-TERS will lead to a better understanding of electron-transport processes in molecular junctions (*41, 42*).

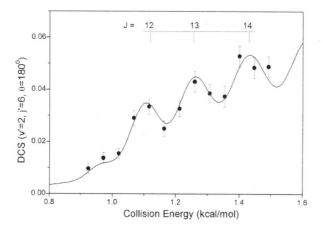

Figure 14. Experimental and theoretical DCS of the HF(v'= 2, j'= 6)product of the F($^2P_{3/2}$) + HD(j = 0) reaction in the backward scattering direction. The solid circles are experimental data; the red curve, the result of full quantum dynamics calculations convoluted with the experimental resolution and shifted 0.03 kcal/mol lower in energy. (see color insert)

Song Gao of Peking University has discovered novel magnetic relaxation phenomena in single-chain magnet and single-ion magnet. By choosing metal ions with large magnetic anisotropy, and by employing azido to transfer magnetic interaction and proper terminal ligands to isolate the individual magnetic chains so as to satisfy Glauber's dynamic conditions, his group has successfully obtained the first examples of homo-spin single-chain magnet (*43*), see Figure 16. Using diamagnetic ions and/or long bridging ligands to separate the paramagnetic ions efficiently, his group has found that some weakly coupled systems exhibit the external field-dependent magnetic relaxation phenomena, which are quite similar to, but different from, the relaxation in superparamagnets . He also discovered a series of novel molecular magnets based on cyanide (CN−), azide (N3−), dicyanide (C(CN)2−), cyanamide (NCNH−) and formate (HCOO−). In particular, he and his collaborators have synthesized some new hetero-metallic (3d-4f,

123

3d-3d', 3d-4d) magnets, mixed-bridged hybrid magnets, weak ferromagnets constructed by asymmetric three-atom single bridges and porous magnets (*44*). These discoveries are of great importance in understanding the quantum origin of molecular magnetism.

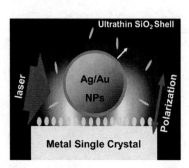 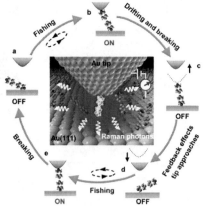

Figure 15. Left: The Principle of shell-isolated nanoparticle-enhanced Raman spectroscopy to obtain surface Raman signal on surfaces or species that do not support enhancement. Right: A schematic diagram of fishing-mode tip-enhanced Raman spectroscopy that allows the Raman and electric conductance signals of single molecules to be obtained simultaneously. (see color insert)

Prof. Wenjian Liu from Peking University developed a method to eliminate symmetrically the small component of the wavefunction in relativistic quantum chemistry, resulting in a new Hamiltonian, that is easily solved. His approach is conceptually simple, numerically accurate, computationally efficient and serves as a seamless bridge between Dirac and Schrödinger equations (*45*). Another contribution of Prof. Wenjian Liu is the novel formulation of relativistic theory for the determination of nuclear magnetic resonance (NMR) parameters. In addition, he has explored the first time-dependent relativistic density functional theory for electronic excitations in heavy atoms. These developments have rendered the BDF (Beijing Density Functional, of which Liu is an important developer) package very powerful in first-principles electronic structure calculations for heavy atoms.

Prof. Shuhua Li in Nanjing University has developed an efficient linear scaling method for computing the electronic structures of large molecules. One of the strategies adopted is the cluster-in-molecule (CIM) local correlation technique, which can extend the applications of *ab initio* correlation methods (CCSD or MP2) to large molecules (*46, 47*). The CIM approach is based on the "locality" of electron correlation, and is considered to be one of the representative linear scaling coupled cluster algorithms. Another strategy is the energy-based fragmentation approach. In this scheme, the total energy of a large molecule can be calculated approximately from energy calculations on a small subsystem. This method has been applied to various biological molecules or molecular clusters at various theoretical levels. It has been used for the calculation of the ground-state energies and structure optimization, vibrational spectra, and some molecular properties at *ab initio* levels for more than 1000 heavy atoms, which are well beyond the reach of traditional computational methods.

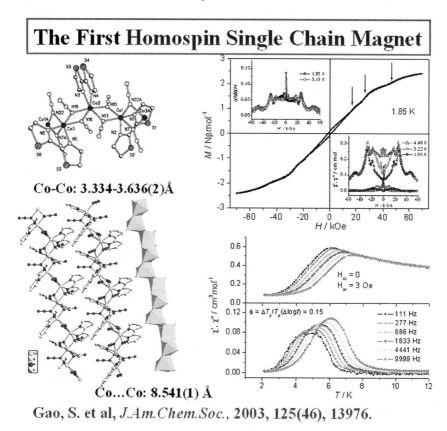

The First Homospin Single Chain Magnet

Co-Co: 3.334-3.636(2)Å

Co...Co: 8.541(1) Å

Gao, S. et al, *J.Am.Chem.Soc.*, 2003, 125(46), 13976.

Figure 16. The first homospin single chain magnet. Reproduced with permission from reference (43). Copyright 2003 American Chemical Society. (see color insert)

It is widely noted that excited state is at the center of interest for theoretical chemistry, both from the electronic structure and from the chemical dynamics perspectives. For complex system, time dependent density functional theory (TDDFT) nowadays becomes the standard computational tool to reveal the electronic excited state structures. Energy gradient is pivotal both for geometry optimization, response theory, as well as property evaluation. While algorithm for the first-order analytical gradient for the excited state in the framework of TDDFT has been obtained a while ago, the second-order gradient has not been available for long time. People rely on numerical evaluation that is both time-consuming and inaccurate. Prof. Wanzhen Liang from Xiamen University has recently been successful in achieving an analytical solution for the second-order gradient with respect to both atomic coordinates and external field. This is a breakthrough in TDDFT with wide application both for excited state structure and chemical dynamics calculations (*48, 49*).

Theoretical chemistry is popularly classified into three domains: structure theory, chemical dynamics, and statistical mechanics. Prof. Zhigang Shuai aims at developing reliable theoretical methods for the prediction of the properties of materials, which requires a trio of the three instead of a solo. The first example is his recent work on quantitative prediction of quantum efficiency of molecular fluorescence. Prof. Zhigang Shuai and Jiushu Shao have recently developed a new analytical formalism for determining the nonradiative decay rate constants by taking into account multimode mixing due to Duschinsky rotation for potential energy surfaces of the excited and the ground states (50). The theory successfully demonstrates completely different photophysical behaviors for the two isomers of tetraphenylbutadienes, thus providing a reasonable explanation of the aggregation induced emission phenomena for one isomer. This allows an "*ab initio*" molecular fluorescence efficiency prediction (51). Shuai *et al.* further employed hybrid quantum mechanics/molecular mechanics (QM/MM) approach to investigate the excited state decay in molecular aggregate (Figure 17). In contrast to the conventional aggregation quenching, some luminophors can exhibit exotic aggregation induced emission phenomena. From Shuai' calculations, it is found that the intermolecular electrostatic interaction could suppress low-frequency motion to reduce the non-radiative decay and to enhance the luminescence quantum efficiency. To predict the charge mobility in optoelectronic materials, Shuai *et al.* have developed a first-principles scheme based on hopping mechanism by combining quantum chemical calculations plus a statistical simulation (Random Walk) to investigate charge transports. He showed that (i) the quantum nuclear tunneling effect is essential, and (ii) the dynamic disorder does not play an appreciable role for charge hopping transport in organic semiconductors. Later, his model was not only found to be able to explain the apparently "paradoxical" experimental observations that optical measurement indicated "localized charge" while electrical measurement indicated "bandlike," but also directly got several experimental verifications (52, 53). Such nuclear tunneling model was even adopted to conducting polymers for clarifying the recent dispute over the charge transport mechanism.

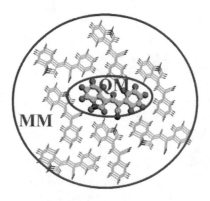

Figure 17. Shuai developed a QM/MM method for studying the quantum efficiency of the organic light-emitting materials.

Sharing International Responsibilities

Chinese chemists are becoming more and more active in the international arena, not only because they are becoming known internationally, but also because they are sharing more and more international responsibilities. We are facing serious global challenges such as environment and natural resources. These are the problems to be solved through chemistry R&D, and international collaborations are indispensable. Under such circumstances, national chemical societies along with corresponding funding agencies from China, US, Japan, Germany and UK have decided to launch a yearly summit to discuss the urgent issues addressing global challenges in energy, environment, resources and water, health and nutrition, and global warming. In 2008, the representatives from the five countries formed a steering committee for the summit, named "Chemical Sciences and Society Symposium (CS3)" and later changed to "Chemical Sciences and Society Summit (CS3)". The purpose is to unite the talent of international chemists to identify the essential challenges in chemistry in order to provide a roadmap to solve these global problems.

The first CS3 took place in Kloster Seeon, Germany, 2009, which was organized by the German Chemical Society, focusing on the subject of "Using Sunlight to Power the World". The major suggestions are shown below:

(i) *Converting solar energy into chemical fuel through artificial photosynthesis mimicking natural processes.* Chemists should develop chemical catalysts for water splitting and CO_2 reduction that can be applied commercially and are made of affordable and earth-abundant materials. Chemists also must create an "artificial leaf" by coupling water splitting and CO_2 reduction in a way that eliminates the need for an external and sacrificial electron donor.

(ii) *Accessing solar energy that already exists in nature.* Chemists must develop biochemical methods that can be used to create more biomass and develop catalytic processes that improve the efficiency of biomass conversion.

(iii) *Converting solar energy into electricity.* Since the widespread use of silicon-based photovoltaic cells is limited due to high cost, chemists must develop low-cost, non-toxic, earth-abundant photovoltaic materials for use in next generation photovoltaic cells.

(iv) *Storing newly harnessed and converted solar energy.* Chemists must develop new catalysts and materials from low-cost, earth-abundant elements that can be used to build affordable, sustainable solar energy transformation and storage systems.

The second CS3 took place in London, organized by the Royal Society of Chemistry in 2010. It focused on the various facets of "Sustainable Materials". It was suggested that materials chemists

(i) must develop new, sustainable energy conversion and storage technologies that can meet the future energy demands, and without increasing harmful emissions of CO_2.

(ii) can help to reduce CO_2 emissions by improving carbon capture and storage systems and developing novel ways of activating and using CO_2 as a value product for fuels and chemical feedstocks rather than waste.

(iii) must develop methods to efficiently obtain petroleum from low-quality sources and processes to efficiently and sustainably utilize fossil fuel alternatives to reduce our dependency on fossil fuel and feedstocks.

(iv) must reduce, replace and recycle the use of scarce natural resources in many applications, as well as developing alternative new materials based on earth-abundant elements.

(v) must use the principles of green chemistry to meet our energy, materials and water needs in ways that are non-harmful and sustainable. New technologies can be developed to better monitor and remove air, soil, and water pollutants from the environment.

The third "CS3" took place in Beijing and was organized by the Chinese Chemical Society in 2011. The summit focused on the theme of "Health". Deadly infectious diseases have been conquered in most regions of the world, but they remain as constant threats. The non-infectious diseases, often chronic, become major concerns. Examples are cardiovascular disease and cancer, which are now the main causes of death. It was recommended that:

(i) Chemists should better understand the chemistry of infection and non-infectious diseases as well as the immune system. They must study the role of reactive oxygen species in age-related diseases through the investigation of brain chemistry and aging processes. Accordingly, novel tools should be developed for studying the life processes.

(ii) Chemists play an essential role in drug design. They should better understand the interaction between drug molecules and biological molecules and develop more sensitive methods for monitoring the interactions between individual molecules in a cell. Chemists must develop better ways to screen natural products and develop new antibiotics and find ways to make drug manufacturing cheaper, more efficient, less wasteful and less reliant on petroleum.

(iii) Chemists must develop better tools for validating and studying biomarkers as well as for detecting individual molecules in cells. They must develop better imaging technologies at a wide range of scales and low-cost genome sequencing technologies.

The fourth "CS3" took place in San Francisco, 2012, organized by the American Chemical Society, focusing on organic and carbon electronics. It was noted that organic electronic devices will do things that silicon-based electronics cannot do, expanding the functionality and accessibility of electronics and will be more energy-efficient and eco-friendly. The following strategies have been recommended:

(i) improving controlled self-assembly. Chemists need to gain better control over the self-assembly of organic electronic molecules into ordered patterns to ensure that the structures being assembled are reproducible, which requires a better understanding of the interfacial behaviors.

(ii) improving three-dimensional processing technology. Chemists must develop processes to fabricate three-dimensional organic electronic structures with the same precision achievable nowadays with two-dimensional printing technology for reliable high-throughput manufacturing of organic electronic devices.

(iii) increasing multi-functionality of organic electronic devices. As chemists gain better control over the synthesis of organic materials, they will be able to build increasingly sophisticated optoelectronic and other devices. Chemists need to broaden their research focus beyond charge carrier transport and gain better understanding of optical, magnetic, thermal and other properties.

(iv) developing better analytical tools. Chemists need better tools for analyzing the molecular composition, molecular organization, and local electronic and other properties of the organic electronic systems.

Chinese Chemical Society (CCS) is a national adhering organization of the International Union of Pure and Applied Chemistry (IUPAC) since 1979 and is also an active member of the Federation of Asian Chemical Societies (FACS). CCS has officially established close cooperation relationships with several national chemical societies, including the Royal Society of Chemistry (RSC), the German Chemical Society (GDCh), the American Chemical Society (ACS), and the Chemical Society of Japan (CSJ). In 2005, ACS sent a delegation with a large number of high profile leaders and chemists to visit CCS, Chinese Academy of Sciences, Ministry of Science and Technology, National Natural Science Foundation, as well as the several elite institutions in chemistry. Since then, CCS-ACS entered into a strategic alliance relationship. CCS-ACS alliance is committed to working together to create a better environment for chemistry to make a positive impact on our society. Every year, the Presidents from both societies issue a joint comment publicly to suggest chemistry solutions to one global issue, as well as to publicize the joint activities of both societies. These comments are published in both "C&E News" and "Chemistry Letters". CCS delegations have paid visits to the headquarters of RSC and GDCh in 2007 to strengthen the established ties. During the CCS Biennial Congress, the leaders from international chemical societies have been invited to come to China for official visits, during which bilateral symposia on mutually interested topic were organized.

Challenges and Opportunities for Chemistry in China

Chemical sciences will continue to be pivotal and indispensable in creating new matters and shaping the development of novel methodologies and new theories. The research frontiers lie in a huge arena ranging from single atoms, single molecules to molecular assemblies in a multiscale world, aiming at establishing the relationship between the properties or functions of these atomic buildups and their structures at different scales. While chemical sciences focus on the unified description at both microscopic and macroscopic levels, revealing the static and dynamic behaviors which will deepen our understandings of matters from simple to complicated perspectives and/or the other way around. The interdisciplinary research with physics, life sciences, materials sciences and information science is essential to solve important problems in environments, energy and resources. Chemical sciences will advance not only in the fundamental understanding of matter, its structure and properties, but also meet the needs of people for progress and prosperity.

Economy is and will still be booming in the foreseeable future in China. There are tremendous opportunities for scientists to carry out cutting-edge research in chemical sciences. Of the most importance includes the exploration of diversity of structures of matter and molecules, synthetic methodologies for creating small molecules, polymer, biomolecules, as well as rational design of the assembly of supramolecular systems, combinatorial, novel synthetic strategy. The structure and transformation of matter are always at the heart of chemistry. It is expected that elucidation of the multiscale structure of matter and investigation on the relationships between structures and properties and functions will be the mainframe of chemical sciences. These structures here generally refer to molecular geometry, the chain conformations of polymers, the folding state, configuration of the supramolecular assembly, higher level structure of biomolecules and formation of nanostructures. The following areas are full of opportunities and challenges.

1. Novel Synthetic Strategies, Concepts, and Methods

As the major means of creating new substances, the methodologies and strategies of inorganic, organic, polymer, solid and combinatorial syntheses lie in the core of chemical sciences. Research in this direction is highly target-oriented, requiring originality, creativity, and interdisciplinary interactions.

The key issue is to develop new synthetic methods, to design and to synthesize compounds of desired properties and new molecular systems, which cannot be realized by merely resorting to trial and error or by traditional procedures. A designing scheme based on structure, mechanism as well as the reaction dynamics should be rationalized. The ultimate goal of modern synthetic chemistry is to make a substance of specified structure and function with high efficiency and selectivity, meeting the needs of functionalized chemicals in life, material, information, energy and environmental sciences. Nowadays, the most active and exciting research frontiers include asymmetric organic synthesis and controllable formation of ordered structures. Novel ideas are required to achieve this goal. A breakthrough in this research will not only help understand the origin of biological homochirality, but also be invaluable in improving the techniques in pharmaceutical, agrichemical, and materials industry.

The advances in synthetic chemistry will also gain momentum from other academic disciplines such as biology, physics, material science, information sciences as well as environmental science. Synthetic strategies by virtue of biological and physical laws and effects can be anticipated to assist chemists to invent new routes for synthesis.

Major challenges include:

(a) Function oriented synthesis: methods and strategy
(b) Highly selective synthesis *a la* catalyzed asymmetric synthesis
(c) Molecular design and synthesis of specific structures and properties
(d) Synthesis and manufacture under mild and extreme conditions
(e) Activation and transformation of inert chemical bonds

2. Chemical Dynamics and Control: Experiments and Theories

Chemical dynamics experiments aim at observing and controlling the real-time evolution of chemical processes, while theories can predict the dynamic behavior and tell the optimal conditions for active control. The dynamics of typical elementary reactions, particularly those involving quantum effects including oscillation of wave packets, tunneling as well as interference can be further explored by experimental measurements and theoretical calculations. The recognition, reaction, and manipulation of single molecules will be a hot topic. The underlying basis of chemical bonding and interaction for single molecules is yet to be clarified. Novel methods in quantum chemistry and chemical dynamics will be developed for more accurate description of structures as well as dynamic processes. The focuses will be the energetics of transient intermediates, reaction pathways, and chemistry and physics of single molecules.

Major challenges include:

(a) Chemical dynamics: Theory and experimental techniques
(b) Understanding and control of ultrafast processes in chemical reactions
(c) Probe of structures and dynamic motion of single molecules on surfaces
(d) Investigation of mechanisms of catalytic reactions
(e) Dynamics and control of chemical reactions on interfaces

3. Molecular Assembly, Ordered Structure, and Function

The properties of matter depend not only on the structure of its building molecules, but to some extent also on the structure of the assemblies. For small molecules, the study on the formation of their crystals and nanostructures will be a fundamental one. For synthetic polymers, investigation on the structure and motion of the polymeric chain as well as diversified condensed states will be essential to design specific structures and new materials that possess desired properties. Once their ordered structures and chemistry of assembly are known, supramolecular systems or nonbonded assemblies can be constructed with inorganic and organic molecules, biomolecules, and polymers. For theorists, a methodology based on chemistry, life sciences, physics for assembly chemistry, full of explanatory and predictive power, is yet to be set up. The theory should include a good description of the nature of the weak intermolecular interaction, electron transfer, energy transfer, matter transport, and chemical transformations. With these, one can establish the relationship between the formation of molecular assembly with specific structure and ordered higher-level structure. Therefore, one can use theoretical findings to design basic units of desired properties and functions.

Major challenges include:

(a) Nature of the weak intermolecular interaction and position recognition
(b) Growth dynamics for molecular assembly and ordered higher structures
(c) Reactivity of relation-structure relationship of molecular assemblies
(d) Construction and functions of novel supramolecular systems

4. Theoretical and Computational Methods for Complex Chemical Systems

Chemical theory will play an increasingly important role in the development of chemical sciences in the 21st century. Future development will focus on the theoretical and computational approaches for complex chemical systems that are important in biology, materials, and environmental sciences. Inorganic solids, large organic and biomolecules, supramolecular systems, interfaces are all examples of complex systems. Not only the characteristics of the microscopic state or gas phase at zero temperature, but also the real property of molecular ensemble or thermal state should be calculated. Theoretical and computational chemistry is essentially an interdisciplinary research area on multi-scale phenomena in both space and time. As an indispensable part of chemistry, in either conceptual understanding or quantitative characterization, theoretical chemistry will continue to benefit from assimilation of state-of-the-art results of mathematics, physics, computer sciences, and other sciences.

Major challenges include:

(a) Computational chemistry
(b) Methodology of assessment of properties based on electronic structures
(c) Theoretical dynamics for micro, meso- and macromolecular systems
(d) Theoretical treatment for the chemical dynamics of interfaces, metastable systems and small systems

5. New Techniques and Approaches for Chemical Analysis and Detection

Analysis, detection, and characterization are the means of acquiring the information on chemical components, structures, and interactions in molecular systems. Advancement in chemical sciences heavily relies on the development of new principles and methodologies for analysis and detection, so do many other fields such as life, material, environments, energy, health and medical sciences, security, and economics. In fact, major advancements in scientific research can be attributed to the improvements in methodologies and tools. For example, the success of the well-known human genome project is primarily due to the establishment of rapid DNA sequencing techniques. The range of targets for analysis, detection and characterization has been extended from simple systems to complex ones. Complex systems consist of many different substances appearing in diversified morphologies, and some components are very tiny and they interact with each other nonlinearly. Advanced analysis methods, sensitive detection techniques, and perfect instrumentation are necessary to carry out qualitative and quantitative analysis in situ and transiently. They should be effective for analyzing huge data, detecting trace substances, and probing the biological activity.

Major challenges include:

(a) Separation and characterization of complex systems
(b) New analysis and detection tools in multiple dimensions, scales and parameters
(c) New methods and technologies in combinatorial chemistry
(d) Theory and methods for interaction and signal transformation of substances
(e) Analysis and detection methods and technologies for the national security and human health

6. Chemical Processes in Life System and Regulation of Function

Chemical biology, a term coined by a Harvard scientist Schreiber, represents an active field nowadays of understanding biology in light of chemistry. This area is an extension and a unification of biological organic, inorganic, structural, and natural product chemistry. It aims at exploring the biological effects and regulation of functions at molecular level. A central subject in the field is the selection from natural or synthetic small compounds such molecules that can regulate the genomic functions, signal transduction, and protein-cell interactions in biological systems and processes. These molecules work as probes and tools for studying the mechanisms of their interaction and recognition with targets in biological networks. Chemical biologists are also to discover the general rules of biological synthesis in nature. By doing so, they can devise effective, combinatorial strategies to synthesize more diversified molecules that may be good precursors for new drugs. Another important subject is to develop reliable techniques for analyzing complex biological systems either statically or dynamically. Fundamentally, chemical biology can help reveal the interaction of large molecules and information transfer in life systems at molecular level.

Major challenges include:

(a) Functional regulation of known biomolecules and their networks with small molecules
(b) Discovering small bioactive molecules and exploring their mechanisms of interaction with biological targets

(c) Improving methods and technologies for acquiring information in biological processes

(d) Discovering new biological targets and networks based on small-molecule probes

(e) Explaining the chemical nature of life processes

7. Fundamental Questions in Green and Environmental Chemistry

Environmental chemists mainly focus on the measurements and control of the existence, characterization, behavior, and effects of chemical substances in different environmental media. Environmental chemistry is typically an interdisciplinary field, which is different from traditional chemistry in the scale of time and space. Understanding the chemistry in environmental sciences is essential to solving the pollution problems. Because environmental pollution aggravates with rapid industrialization, Chinese chemists should contribute more to protect the natural surroundings and ecologic systems.

As indicated earlier, environmental chemistry plays important roles in monitoring, detecting, categorizing the sources of pollution, understanding the related chemical and ecological processes and effects, and controlling as well as preventing chemical pollutions. However, to entirely get rid of the environmental problems, one has to resort to green chemistry. The mission of green chemistry is to "promote innovative chemical technologies that reduce or eliminate the use or generation of hazardous substance in the design, manufacture, and use of chemical products" (see the website of U.S. Environmental Protection Agency). Aiming at sustainable development, green chemists perform the cutting-edge research for manufacturing safer products and reducing waste, energy, and resources. The final goal is to develop environmentally friendly means of chemical production. Green chemistry will eventually result in a revolutionary change in chemical sciences and should be a sound basis for a sustainable and harmonious development of modern society.

Major challenges in green chemistry include:

(a) Highly effective and selective reactions and processes with atomic economy

(b) Development and use of environmentally friendly reaction media

(c) Transformation of use of recycled materials, biodegradable materials and biomass

(d) Integration of processes based on green chemistry

Major challenges in environmental chemistry include:

(a) Automatic, online, in situ, dynamic, and transient analysis of pollution

(b) Understanding and predicting transport rule, ecological risk, and health effects of persistent poisonous and hazardous substances

(c) Fundamental principles and methods for controlling pollution

(d) Environmental molecular sciences

(e) Applications of biological, information, and material techniques in environmental chemistry

8. Chemistry of Material Science

Materials, information and energy are the indispensable tripod for modern civilization. Certainly, materials are essential for humankind to survive and to exploit. Molecular design, synthesis, elucidation of structure and properties of materials, and mechanisms of surface reactions are all core subjects in traditional chemical sciences. Chemistry is no doubt the cradle of new materials. The last two decades witnessed much progress in material science, exemplified by the discovery of numerous functionalized molecules. They include electro-optical and magnetic polymers, fullerenes and carbon nanotubes, self-assembled monolayers, ferrocene catalysts, combinatorial materials, nanocrystals, and various industrial plastics, rubbers and fibers. Computer simulation of the structures and properties of complex materials as well as computer-assisted design becomes a powerful tool in material science.

Major challenges in chemistry of materials include:

(a) Relationships between the chemical structure-assembly morphology and properties
(b) Numerical simulation, syntheses, and synergistic effects of smart composite materials
(c) Design, pollution-free syntheses and stability of environmental friendly and biocompatible materials
(d) Control of the chemical synthesis and morphology under mild or extreme conditions
(e) Chemistry of failure and remediation processes of materials

9. Fundamental Chemistry in Energy and Resources

Manifesting a nation's comprehensive strength as well as standard of living and cultural achievements, energy and resources are the material basis for economic and social development. Nowadays, energy resources are so critical for sustainable development that the energy systems established last century no longer meet the current needs for high efficiency, cleanliness, economy, and safety. In China, there are many opportunities in the research on chemistry of coal, gas, and the transformation between coal and gas. To develop cutting-edge techniques in this and other resources such as solar, biological, and hydrogen energy is of high priority.

Major challenges include:

(a) Chemistry of effective, clean refining of fossil fuels
(b) Control and conversion of greenhouse gases
(c) Renewable and clean energies

(d) Chemistry of highly effective, clean battery

(e) Physical chemistry and utilization of specific resources in China

10. Essential Scientific Questions in Chemical Engineering

Emerging from applied chemistry, chemical engineering is an academic discipline on mass transport, energy transfer, and other related processes in manufacturing chemical products. Its goal is to improve the quality of chemical industry with the expertise from chemistry, physics, and mathematics. Chemical engineers focus on the study of the dependence of reactions on mass transport in industry. They develop new approaches for amplification, and control of matter transformation, and the design of equipments for effective, energy saving, economy, and safe production.

Major challenges include:

(a) Fundamental theory of chemical engineering for large-scale manufacturing and applications of advanced materials

(b) Chemical reaction, biological transformation and ultra-large separation processes

(c) Numerical simulation and information acquisition, retrieval, and applications of chemical engineering processes

(d) Control of the complex structures in chemical engineering

Perspective

Chemistry has indeed contributed largely to the social development and civilization. It can be seen in every corner of human activities, from plastics and daily consumables, food and nutrition, drugs and genome project to energy and transport, materials and information technology. Although chemistry has tremendously profited the modern society, unfortunately its image has been tarnished by the negative publicity associated with pollution of chemical industry, green house effects, and other environmental problems. It is the scientists' mission to build a consensus with the society that these negative effects can be remedied only through the advancements in science. Thus, general chemical education is a necessary responsibility for all chemists worldwide. We should tell people that chemistry, through further advancements instead of retrogradations, can provide solutions to problems ranging from global warming, pollution, resources limitation to food shortage and fatal diseases and epidemics. In summary, chemistry is one of the key scientific foundations for the sustainable development of a harmonious society.

Acknowledgments

We take this opportunity to express our sincere gratitude towards many colleagues for constructive suggestion and insightful comments. We are indebted to Prof. Chunli Bai, the President of Chinese Academy of Sciences and the Past President of the Chinese Chemical Society, for sharing his broad vision on chemistry with us. He proof-read and made corrections on the first draft of this article. We are especially indebted to Prof. Jiannian Yao, President of the Chinese Chemical Society and Vice-President of the National Natural Science Foundation of China, for his constant support in international activities of CCS and for sharing his opinions on the advances of chemical sciences in China. The authors also thank Prof. Wenping Liang and Yongjun Chen, Directors of the Department of Chemical Sciences of the National Natural Science Foundation of China (NSFC), for providing the General Outline of Chemical Sciences Development in China. The following scientists have provided invaluable research highlight (alphabetic order): Chunli Bai, Yong Cao, Song Gao, Maochun Hong, Guibin Jiang, Shuhua Li, Yadong Li, Wanzhen Liang, Wenjian Liu, Shengming Ma, Yanlin Song, Shigang Sun, Zhongqun Tian, Chen-Ho Tung, Lijun Wan, Deyue Yan, Jinlong Yang, Xueming Yang, Jiannian Yao, Dongyuan Zhao, Jincai Zhao, Lansun Zheng and Qilin Zhou. We would like to apologize any inaccurate presentation of other's work. The authors are indebted to Dr. Huaping Xu for careful proofreading of the manuscript and making corrections and adjustments of the tables, schemes, figures, and references.

References

1. Bai, C. L. *Chin. Sci. Bull.* **2009**, *54*, 1941.
2. Zhou, H. H.; Song, Y. L. *Adv. Mater. Res.* **2011**, *174*, 447.
3. Yao, X.; Song, Y. L.; Jiang, L. *Adv. Mater.* **2011**, *23*, 683.
4. Tian, D. L.; Song, Y. L.; Jiang, L. *Chem. Soc. Rev.* **2013**, *42*, 5184.
5. Zhang, Z. L.; Zhang, X. Y.; Xin, Z. Q.; Deng, M. M.; Wen, Y. Q.; Song, Y. L. *Nanotech.* **2011**, *22*, 425601.
6. Zhao, A. D.; Li, Q. X.; Chen, L.; Xiang, H. J.; Wang, W. H.; Pan, S.; Wang, B.; Xiao, X. D.; Yang, J. L.; Hou, J. G.; Zhu, Q. S. *Science* **2005**, *309*, 1542.
7. Wang, X.; Zhuang, J.; Peng, Q.; Li, Y. D. *Nature* **2005**, *437*, 121.
8. Chen, T.; Yang, W. H.; Wang, D.; Wan, L. J. *Nat. Commun.* **2013**, *4*, 1389.
9. Xie, S. Y.; Gao, F.; Huang, R. B.; Wang, C. R.; Zhang, X.; Liu, M. L.; Deng, S. L.; Zheng, L. S. *Science* **2004**, *304*, 699.
10. Tian, N.; Zhou, Z. Y.; Sun, S. G.; Ding, Y.; Wang, Z. L. *Science* **2007**, *316*, 732.
11. Chen, H. J.; Guo, Y. L.; Yu, G.; Zhao, Y.; Zhang, J.; Gao, D.; Liu, H. T.; Liu, Y. Q. *Adv. Mater.* **2012**, *24*, 4618.
12. Zhang, J.; Geng, H.; Virk, T. S.; Zhao, Y.; Tan, J. H.; Di, C. A.; Xu, W.; Singh, K.; Hu, W. P.; Shuai, Z. G.; Liu, Y. Q.; Zhu, D. B. *Adv. Mater.* **2012**, *24*, 2603.

13. Zhang, F. J.; Di, C. A.; Berdunov, N.; Hu, Y. Y.; Hu, Y. B.; Gao, X. K.; Meng, Q.; Sirringhaus, H.; Zhu, D. B. *Adv. Mater.* **2013**, *25*, 1401.

14. Feng, F. D.; Liu, L. B.; Wang, S. *Nat. Protoc.* **2010**, *5*, 1255.

15. Xing, C. F.; Xu, Q. L.; Tang, H. W.; Liu, L. B.; Wang, S. *J. Am. Chem. Soc.* **2009**, *131*, 13117.

16. He, Y. J.; Chen, H. Y.; Hou, J. H.; Li, Y. F. *J. Am. Chem. Soc.* **2010**, *132*, 1377.

17. Huang, F.; Hou, L. T.; Wu, H. B.; Wang, X. H.; Shen, H. L.; Cao, W.; Yang, W.; Cao, Y. *J. Am. Chem. Soc.* **2004**, *126*, 9845.

18. Zeng, W. J.; Wu, H. B.; Zhang, C.; Huang, F.; Peng, J. B.; Yang, W.; Cao, Y. *Adv. Mater.* **2007**, *19*, 810.

19. He, Z. C.; Zhong, C. M.; Su, S. J.; Xu, M.; Wu, H. B.; Cao, Y. *Nat. Photon.* **2012**, *6*, 591.

20. Zhang, C.; Yan, J.; Yao, J. N.; Zhao, Y. S. *Adv. Mater.* **2013**, *25*, 2854.

21. Lei, Y.; Liao, Q.; Fu, H. B.; Yao, J. N. *J. Am. Chem. Soc.* **2010**, *132*, 1742.

22. Wan, Y.; Zhao, D. Y. *Chem. Rev.* **2007**, *107*, 2821.

23. Yan, D. Y.; Zhou, Y. F.; Hou, J. *Science* **2004**, *303*, 65.

24. Gao, C.; Yan, D. Y. *Prog. Polym. Sci.* **2004**, *29*, 183.

25. Wang, C.; Wang, Z. Q.; Zhang, X. *Acc. Chem. Res.* **2012**, *45*, 608.

26. Wang, Y. P.; Ma, N.; Wang, Z. Q.; Zhang, X. *Angew. Chem., Int. Ed.* **2007**, *46*, 2823.

27. Wang, C.; Ying, S. C.; Xu, H. P.; Wang, Z. Q.; Zhang, X. *Angew. Chem., Int. Ed.* **2008**, *47*, 9049.

28. Li, H. R.; Wu, L. Z.; Tung, C. H. *J. Am. Chem. Soc.* **2000**, *122*, 2446.

29. Tung, C. H.; Wu, L. Z.; Zhang, L. P.; Chen, B. *Acc. Chem. Res.* **2003**, *36*, 39.

30. Feng, K.; Zhang, R. Y.; Wu, L. Z.; Tu, B.; Peng, M. L.; Zhang, L. P.; Zhao, D. Y.; Tung, C. H. *J. Am. Chem. Soc.* **2006**, *128*, 14685.

31. Xie, J. H.; Zhou, Q. L. *Acc. Chem. Res.* **2008**, *41*, 581.

32. Zhu, S. F.; Zhou, Q. L. *Acc. Chem. Res.* **2012**, *45*, 1365.

33. Li, S.; Zhu, S. F.; Zhang, C. M.; Song, S.; Zhou, Q. L. *J. Am. Chem. Soc.* **2008**, *130*, 8584.

34. Xie, J. H.; Liu, X. Y.; Xie, J. B.; Wang, L. X.; Zhou, Q. L. *Angew. Chem., Int. Ed.* **2011**, *50*, 7329.

35. Zhu, S. F.; Cai, Y.; Mao, H. X.; Xie, J. H.; Zhou, Q. L. *Nat. Chem.* **2010**, *2*, 546.

36. Zhu, B. Z.; Zhao, H. T.; Kalyanaraman, B.; Liu, J.; Shan, G. Q.; Du, Y. G.; Frei, B. *Proc. Natl. Acad. Sci. U.S.A* **.2007**, *104*, 3698.

37. Zhu, B. Z.; Kalyanaraman, B.; Jiang, G. B. *Proc. Natl. Acad. Sci. U.S.A.* **2007**, *104*, 17575.

38. Zhao, W.; Ma, W. H.; Chen, C. C.; Zhao, J. C.; Shuai, Z. G. *J. Am. Chem. Soc.* **2004**, *126*, 4782.

39. Qiu, M. H.; Ren, Z. F.; Che, L.; Dai, D. X.; Harich, S. A.; Wang, X. Y.; Yang, X. M.; Xu, C. X.; Xie, D. Q.; Gustafsson, M.; Skodje, R. T.; Sun, Z. G.; Zhang, D. H. *Science* **2006**, *311*, 1440.

40. Deng, W. H.; Xiao, C. L.; Wang, T.; Dai, D. X.; Yang, X. M.; Zhang, D. H. *Science* **2010**, *327*, 1510.

41. Li, J. F.; Huang, Y. F.; Ding, Y.; Yang, Z. L.; Li, S. B.; Zhou, X. S.; Fan, F. R.; Zhang, W.; Zhou, Z. Y.; Wu, D. Y.; Ren, B.; Wang, Z. L.; Tian, Z. Q. *Nature* **2010**, *464*, 392.

42. Liu, Z.; Ding, S. Y.; Chen, Z. B.; Wang, X.; Tian, J. H.; Anema, J. R.; Zhou, X. S.; Wu, D. Y.; Mao, B. W.; Xu, X.; Ren, B.; Tian, Z. Q. *Nat. Commun.* **2011**, *2*, 305.

43. Liu, T. F.; Fu, D.; Gao, S.; Zhang, Y. Z.; Sun, H. L.; Su, G.; Liu, Y. J. *J. Am. Chem. Soc.* **2003**, *125*, 13976.

44. Wang, X. Y.; Wang, L.; Wang, Z. M.; Gao, S. *J. Am. Chem. Soc.* **2006**, *128*, 674.

45. Liu, W. J. *Mol. Phys.* **2010**, *108*, 1679.

46. Li, S. H.; Li, W.; Fang, T. *J. Am. Chem. Soc.* **2005**, *127*, 7215.

47. Li, S. H.; Shen, J.; Li, W.; Jiang, Y. S. *J. Chem. Phys.* **2006**, *125*, 074109.

48. Liu, J.; Liang, W. Z. *J. Chem. Phys.* **2011**, *135*, 184111.

49. Liu, J.; Liang, W. Z. *J. Chem. Phys.* **2011**, *135*, 014113.

50. Peng, Q.; Yi, Y. P.; Shuai, Z. G.; Shao, J. S. *J. Chem. Phys.* **2007**, *126*, 114302.

51. Peng, Q.; Yi, Y. P.; Shuai, Z. G.; Shao, J. S. *J. Am. Chem. Soc.* **2007**, *129*, 9333.

52. Shuai, Z. G.; Wang, L. J.; Li, Q. K. *Adv. Mater.* **2011**, *23*, 1145.

53. Geng, H.; Peng, Q.; Wang, L. J.; Li, H. J.; Liao, Y.; Ma, Z. Y.; Shuai, Z. G. *Adv. Mater.* **2012**, *24*, 3568.

Chapter 13

Current Status and Role of the Korean Chemical Society in a Changing Korean Society

Han-Young Kang*

President, Korean Chemical Society, Department of Chemistry,
Chungbuk National University, Cheongju 361-763, Republic of Korea
*E-mail: hykang@chungbuk.ac.kr

The Korean Chemical Society (KCS) has undergone rather dramatic growth in terms of socioeconomic and academic performances during the past decades. Currently chemical communities play a key role in various Korean industrial fields, including fine chemicals, polymers, petroleum refining, electronics, biochemicals, medicines, etc, by providing highly trained human resources as well as technical innovation. Continued successful growth of chemical industry has led to the rapid expansion of research activities in the Korean chemical community. Furthermore, KCS has striven to make a significant contribution to the global academic enterprise, and we hope to continue to make changes in the future. However, there are concerns and challenges in the field of chemistry as we continue along the current successful path. Herein, we would like to present the current status of KCS, the expected role of KCS in Korean academia and industry, and our endeavors to publicize chemistry to the public.

Introduction

In the Republic of Korea we have in total about 8,000 professional chemists in both academia and research institutes. In academia alone we have 2700 graduate students, 1,600 in the master degree program and 1,100 students in Ph.D. program (2011). Many different areas of research are being conducted in National Research Institutes, supported by the Korean government. KRICT (Korea Research Institute of Chemical Technology) is especially devoted to chemical research.

KCS was established in 1946 and is the oldest scientific society in Korea. We currently have more than 7000 active members, making KCS one of the largest scientific societies in Korea. There are 12 regional and 13 academic chapters. Two general meetings are held semi-annually, in spring and fall. We accommodate about 3,000 participants with 1,500 presentations (including posters) at each of these meetings. We are now publishing several scientific journals, two of which are associated with VCH–Wiley. We also have a monthly news magazine.

We cooperate with various international organizations such as IUPAC, Federation of Asian Chemical Societies (FACS) and others. KCS is also a member of KUCST (Korean Union of Chemical Science and Technology Societies), the organization of chemistry-related societies. Five chemistry-related societies, Polymer Society of Korea, Korean Society of Industrial and Engineering, Korean Ceramic Society, Korean Institute of Chemical Engineers, and KCS, have worked closely together. KCS is also one of the major members of KOFST (Korea Federation of Science and Technology Societies), which represent the entire community of 500 million scientists and engineers in Korea.

Korean Chemical Industry

The chemical industry in Korea has played an important role in Korea's economic development (Figure 1). The history of South Korea's chemical industry since 1960 comprises four phases. The first three phases, between the 1960s and 1990s, were driven by nationwide economic development plans. Since the 1990s, the private sector has been the main force of the changes.

The Korean chemical industry has made remarkable contributions to the country's economic prosperity (Figure 2). In 2011, Korea's chemical industry was the world's 6th largest in the global chemical market. The chemical industry is the largest in Korean domestic manufacturing. The production comprises 23% of Korea's total manufacturing capacity (2011). It is estimated that chemical raw materials and products stimulate 173 billion USD industrial production. This is the second largest in Korean domestic manufacturing. In terms of trade, Korea exports 77.7 billion USD chemical materials and import 60.9 billion USD, with a net surplus of 16.8 billion USD in 2011.

While Korean chemical industry is strong, there are challenges: 1) deepening trade imbalance in the fine chemical sector, 2) small and medium enterprises that are vulnerable due to weak cooperation with big companies, 3) insufficient R&D investments relative to other industries, 4) petroleum-based and energy-consuming industrial structure, 5) weak cooperation among industry, academia, research institutes, and government, 6) scattered long-term development strategy for the chemical industry, and 7) imbalance in supply and demand in the manpower of the chemical field.

The school age population of those who are entering university-level studies is a matter of concern. As Korea is becoming an aging society, the number of students who are eligible to enter the university reached its possible maximum in 2012. After 2012, the number will start to go down quickly. It is expected that university entrance quota will exceed the number of high school graduates in 2018, which means that we are not going to have enough students in Korean universities. This will result in a severe shortage of high school students.

Figure 3 shows the current status of manpower in chemistry. We have approximately 500,000 people working in chemistry and related areas. In chemical industry, those employed totaled ~346,000 people in 2009 which corresponded to 14% of the entire manpower of the all the manufacturing industries. Technical manpower, which means those who have been educated in college or higher-level institutions reached ~44,000 (12.3% of manpower in all manufacturing industry) in this same time period. We have ~112,000 students in universities including both undergraduate and graduate students. The number of faculty members is approximately 4,000 (2011).

Future chemical industry also expects a shortage of technical personnel. Sectors other than chemical industry (e.g., semiconductor, automobile, and shipbuilding) have been able to attract people to some degree, but it is predicted that chemical industry in Korea will experience a serious imbalance between supply and demand (about 2800 technical people). Since there are about 44,200 technical people in the Korean chemical industry, this number translates into about 6% shortage.

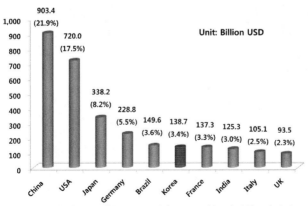

Ranked 6th largest in the global market
-The production scale of Korea's chemical industry in 2011 was $138.7 Billion, 3.4% in the world
-The World Chemical Market: $4.13 Trillion
-The Asian Market: 43.2% (China is the largest)

Source: American Chemistry Council (2011), "Economic data generated from the Guide to the Business of Chemistry"

Figure 1. Production scale of Korean chemical industry. (Adopted from the KRICT report: CHEMI 2020). [Green to Blue CHEMI 2020: Convergent High-value Earth-saving Magic Initiatives; KRICT(Korea Research Institute of Chemical Technology): Daejeon, Korea, December 2012].

Development of Korea's Chemical Industry

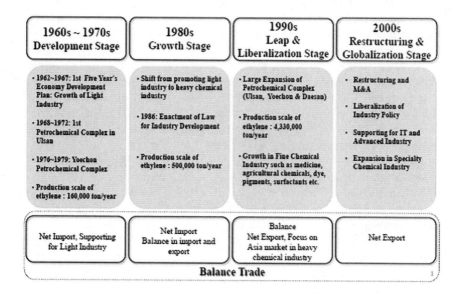

Figure 2. Development of Korean chemical industry. (Adopted from the KRICT report: CHEMI 2020). [Green to Blue CHEMI 2020: Convergent High-value Earth-saving Magic Initiatives; KRICT(Korea Research Institute of Chemical Technology): Daejeon, Korea, December 2012].

Manpower in Chemistry

Total number of manpower in all chemical fields: 500,000

Chemical Industry

- The total number of the employed in chemical industry :
  346,000 persons, 14.1% of the manufacturing industry (2009)
- Technical manpower (over college graduates) :
  44,200 persons which is the 3rd place of 8 industries,
  12.3% of the manufacturing industry (2011)

University (2011)

- Undergrad & Graduate Students : 112,400 (3.4% of total students)
- Faculty : 3,650

Public Research Institute (2011)

- Total 31,700 (Industrial 8,000 Basic 7,500)

Figure 3. Manpower in chemistry in Korea. (Adopted from the KRICT report: CHEMI 2020). [Green to Blue CHEMI 2020: Convergent High-value Earth-saving Magic Initiatives; KRICT(Korea Research Institute of Chemical Technology): Daejeon, Korea, December 2012].

Current Status of R&D in Korea

A summary of the current situation is given in Figure 4. The total research funding is growing and now amounts to 53 billion USD, which is ranked 6th in OECD (Organization of Economic Cooperation and Development) countries. This equals 3.74% of GDP in Korea. Government research funding has also increased, up to 15 billion USD. The number of people working in R&D has increased to ~345,000, which leads to an increase in the number of SCI papers published, up to ~40,000 in 2010 (ranked 11th in the world).

During the recent 10 years, Korean researchers in chemical sciences have published ~60,000 papers, which have ranked the 9th place in the world. In 2011 alone, Korea ranked 6th by publishing 8,931 papers (Figure 5).

Research Funds

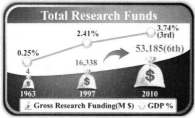

*Source: OECD, Main S&T Indicators(2011/2)

Government Research Funding

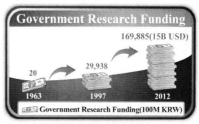

*Source: Nat'l Sci & Tech Commission(2011)

Research Man Power

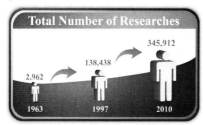

*Source: KISTEP, Report(2011)

Increase in Number of Papers

*Source: Nat'l Sci & Tech Commission(2011)

Figure 4. Current Status of R&D in Korea.

SCI Research Papers ('02~'11)
Recent 10 years (57,587 papers) published (9th); 8,931 papers in 2011 (6th)

Rank	Country	2007	2008	2009	2010	2011	'02~'11
1	USA	25,275	25,933	26,766	28,499	30,090	257,188
2	China	24,059	27,994	31,649	34,560	40,756	240,470
3	Japan	12,780	13,157	12,979	12,326	12,969	133,481
4	Germany	10,086	10,281	10,757	11,536	12,053	103,746
9	Korea	5,676	5,781	6,798	7,571	8,931	57,587
	Total	143,306	151,264	159,000	166,912	182,117	1,439,232

*Source: Web of Science, 2012.9

Research Papers in Collaboration ('02~'11)
High collaboration with USA researchers (9.65%): Coauthorship

country	USA(1)	Japan(2)	China(3)	India(4)	Germany (5)	England (6)	France (7)	Canada (8)
papers	5,566	2,416	1,812	1,100	593	431	386	369
Ratio(%)	9.65	4.20	3.15	1.91	1.03	0.75	0.67	0.64

*Source: Web of Science, 2012.9

SCI papers published in Korea Journals(Chemical Sciences) ('02~'11)
BKCS-most published Journal [recent 10 years (4,717 papers)

Rank	1	2	3	4
Journals	Bulletin of the Korean Chemical Society	Korean Journal of Chemical Engineering	Journal of Industrial and Engineering Chemistry	Macromolecular Research
Papers	4,717	1,579	1,195	990

*Source: Web of Science, 2012.9

SCI papers : Average Citation in 5-yr Intervals ('00~'10)

	'00~'04	'02~'06	'04~'08	'06~'10
OECD	4.44	5.17	5.86	6.45
KOREA	3.05	3.91	4.53	4.97

*Source: National Science Indicators 2011

Figure 5. Summary of SCI chemical research papers published.

Opportunities and Strategies

What are our roles and our current efforts to overcome some of the above challenges? As we know, a new Korean government took office last February. One of the 'hot' terms used by the new government is "creative economy." This has become a key word representing the new paradigm in economic policy that seeks a new way to boost the Korean economy. Simply speaking (although more complicated in reality), creative economy means that with creative ideas, you can generate more intellectual properties, which eventually would lead to more profits that benefit the economy. In order to achieve creative economy, we are required to transform ourselves from "fast follower" to "first mover." To be creative, we have to accomplish the convergence of diverse academic disciplines. In turn, understanding of the basics becomes more important. We believe that the new government will emphasize the importance of basic sciences, and we hope that this will entail an increase in the support of basic sciences.

One of the key questions in this symposium concerns the future trends in chemistry. I think the importance of chemistry in academia in general will keep growing. There will be more and more researchers who are involved with chemistry. Chemistry is considered a central, useful, creative, and cooperative science. As such, we the chemists should make chemistry more readily accessible to other research areas.

One of the issues we are aware of is the need to publicize the benefits of chemistry. In order for chemistry to be more accessible to and benefit the public, communication with the public is critical. It is crucial to make chemical terms and names of compounds less confusing in the Korean language. We are currently updating and organizing chemical terms and nomenclature of compounds in Korean. Communicating these terms to the public is an important task, even though the reward might not be instantaneous.

The second issue concerns a very unique project and a good experiment for publicizing chemistry. KCS has created an award, 'Carbon Culture Award' to acknowledge the efforts of people to bridge the gap between science and other areas, in particular the social sciences. While we have many awards to express our appreciation for scientific achievements, this particular prize is considered very unique and rare for the scientific community. It would offer an excellent opportunity to recognize people in science and those in the humanities.

Carbon is considered responsible for environmental problems from fossil fuels to climate changes. As true as these may be, we should also work on recognizing the potential of its other image as the basis of life. In 2012, we initiated the Carbon Culture Award, and Professor Ynhui Park, a philosopher, was awarded for voicing a world-view in harmony with nature and humans. The Carbon Culture Award has attracted a lot of attention from media and the public. By installing 'Carbon Culture Center', KCS is planning to support the activities of publicizing chemistry to the public.

What would be the needs of chemists? In Korea we really hope for more and easier support for basic sciences. There is a tendency to offer research money to projects according to the 'Research on Demands' policy. However, as genuinely creative research cannot be developed solely by short-term projects with pre-set aims, it is especially important for chemists in Korea to express counter-opinions and work on diversifying the roles and needs of chemistry in real world settings–both in society and in research communities.

Manpower is beginning to be a real problem in this research community. We have to infuse our workforce with creativity, which means that education has to be changed. To enhance the level of research in basic sciences and to meet the demand of manpower, one possible solution is to take the route of globalization. We need to seek supports from not just within Korea but from all over the world.

Conclusions

Currently we are facing many problems and tasks, in regard to energy, environment, health, and sustainable development. The solution is possible only through the combined efforts of people from diverse areas. Science and technology are responsible for tackling technical issues. Chemistry should play a critical role in these efforts due to its characteristics as a central science.

KCS has been paying great attention to these issues and making efforts to activate basic R&D, bring up issues on energy and environment, and emphasizing manpower needs in chemistry. In line with these efforts, KCS has established the Carbon Culture Award and will continue to publicize chemistry through the Carbon Culture Center, an organization under KCS. Furthermore, KCS will increase its involvement with the urgent social issues and national agenda in collaboration with other chemistry-related societies.

ACS is one of the largest scientific societies and has a responsibility and power to participate in the global chemical community, paving the way for other chemical societies in the future. For this reason, this symposium offers KCS an opportunity to examine its collaboration with chemical societies from other countries. KCS will continue to make efforts in globalization by collaborating with foreign societies in the area of education and R&D, and sharing our experiences with those who need help.

Chapter 14

Our Endeavors Directed Towards Chemistry-Driven Sustainable Society

Kohei Tamao*

**President, Chemical Society of Japan, and Science Advisor, RIKEN,
2-1 Hirosawa, Wako, Saitama 351-0198, Japan
*E-mail: tamao@riken.jp**

The CSJ focuses on its efforts directed towards chemistry-driven sustainable society by solving many of its current global challenges by chemistry. The March 2011 disasters in Japan have made a great shift of Japan's science and technology policy from discipline-oriented strategy to issue-driven innovation and recovery, and also caused a big loss of public trust in science, which we need to recover through our efforts in promoting basic and cutting-edge research. Under such circumstances, we should keep in mind the following pronouncements: "Chemists for society," "Chemists in society," as well as "Chemists for science." The CSJ published a set of "Overviews of Chemistry Dream Roadmap for 2040" in March 2012 by summarizing many research themes related to environment, energy, health, new materials, and new scientific frontiers. The CSJ also promotes Japan's "Elements Strategy Initiative" proposed by its chemists in 2004 as a world-leading concept. The CSJ has been successfully organizing several international joint symposia together with Asian countries in order to foster young talent. We also actively participate in the CS3 (Chemical Sciences and Society Summit) together with USA, UK, Germany, and China, as well as Pacifichem and Asian Chemical Congress, as opportunities for the global chemistry community to work together.

An Overview of Chemistry in Japan

The motto of the Chemical Society of Japan (CSJ) is "Toward the chemistry-driven sustainable society." CSJ was founded 135 years ago, now with more than 31,000 indivdual members (65% academic, 35% industrial) and 500 corporate members. It has an annual budget of 900M yen (about 10M USD), 7 regional branches, and 21 divisions. Our staff is capably led by Executive Director Nobu Kawashima.

First, I will focus on four main events during the year 2011, which have had a large impact, not only on us but also on the entire science community in Japan. For the purpose of this article, I will skip the activities under IYC.

1. The Great East Japan Earthquake (GEJE, or 3/11 Disaster)
2. The 4th Science and Technology Basic Plan
3. Chemistry Dream Roadmap up to 2040
4. International Year of Chemistry (IYC)

It is well known that, Japan suffered a large earthquake on March 11, 2011. A year later, the Japanese governmental agency MEXT (Ministry of Education, Culture, Sports, Science and Technology) published a white paper, called "Towards a robust and resilient society - Lessons from the GEJE." The white paper indicated that the lessons learnt from the GEJE should also be a valuable asset for humanity as a whole. As a damaged nation, establishing the process to overcome various problems and sharing the process with the world are the critical challenges in Japan.

The white paper reported several important analyses on how the GEJE affected the Japanese public's confidence in science and technology (ST). Prior to 3/11, 84.5% of the Japanese public indicated trust in scientists. Right after 3/11, only 40.6% of the Japanese indicated trust in scientists. By February 2012 (about 1 year later), we could see some recovery – 66.5% of the Japanese public indicated trust in scientists.

In another survey, the public was asked if it was better for scientists to determine the R&D direction in ST. Prior to 3/11, 78.8% of the public indicated agreement, but after 3/11 only 45% agreed. These are the situations we are now facing in Japan.

MEXT issued the 4th ST Basic Plan for five years in 2011, several months after the 3/11 disaster (Figure 1). It emphasized an important role for scientific research to realize sustainable growth and reconstruction from the disaster. Science policy and research directions were redirected: from "Science and Technology" (ST) to "Science, Technology and Innovation" (STI); from discipline-oriented to issue-driven or problem-solving; furthermore, research in green science and life science fields was given greater emphasis.

As scientists, we accept these realities. *We know we need to play a central role to overcome various problems and to restore the public's confidence in ST toward a sustainable society.* This is my Message #1.

In order to realize our long-term challenges, we need a scientific road map. The Science Council of Japan has published a Science Roadmap in 2011 and the

CSJ has also published "A Chemistry Dream Road Map up to 2040" in March 2012. The road map is given in Figure 2. Important keywords are roughly arranged outward from the center (from 2012 to 2040) and organized into five important areas: environment, energy, new materials, life, and pioneering new fields. In the upper right side we can find five colors corresponding to the fields of organic, inorganic, bio, physical and nano.

Figure 1. Summary of the 4th MEXT ST Basic Plan.

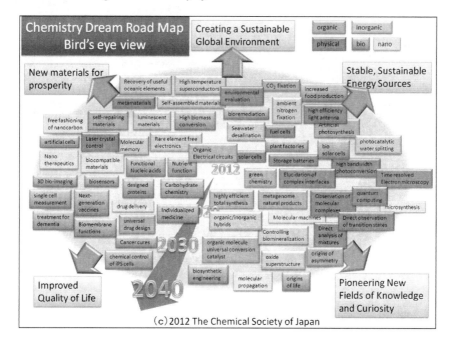

Figure 2. CSJ Chemistry Dream Road Map. (Courtesy of the CSJ). (see color insert)

A close-up of the upper right area of the map is shown in Figure 3. We can see some keywords, such as solar cells, fuel cells, ambient nitrogen fixation, artificial photosynthesis, etc. These would be the challenges for us, as chemists.

We also prepared separate maps for five field, as depicted in Figures 4-8.

Figure 3. Close-up of the right-top part of CSJ Chemistry Dream Road Map. (Courtesy of the CSJ). (see color insert)

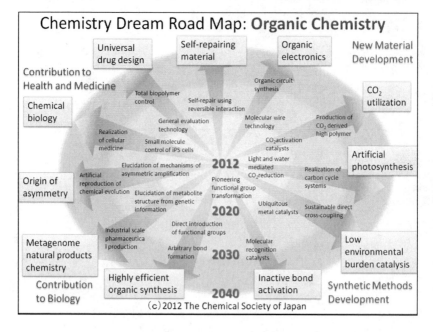

Figure 4. CSJ Chemistry Dream Road Map: Organic Chemistry. (Courtesy of the CSJ). (see color insert)

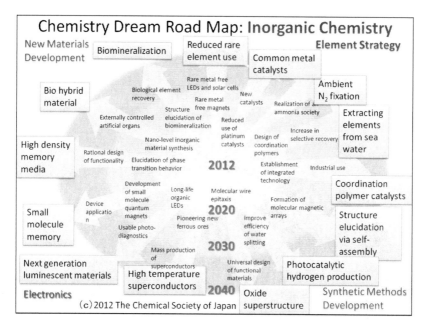

Figure 5. CSJ Chemistry Dream Road Map: Inorganic Chemistry. (Courtesy of the CSJ). (see color insert)

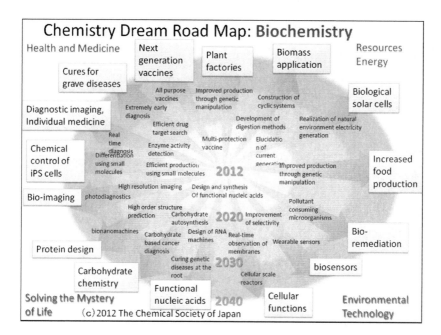

Figure 6. CSJ Chemistry Dream Road Map: Biochemistry. (Courtesy of the CSJ). (see color insert)

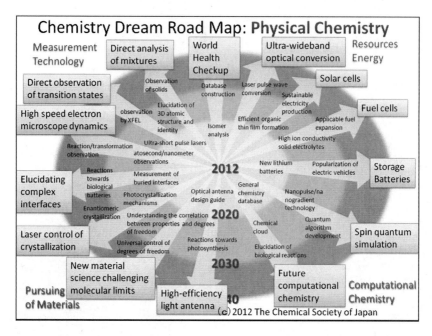

Figure 7. CSJ Chemistry Dream Road Map: Physical Chemistry. (Courtesy of the CSJ). (see color insert)

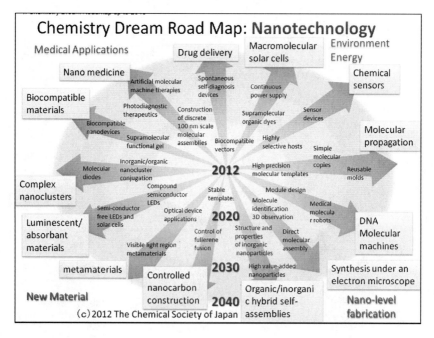

Figure 8. CSJ Chemistry Dream Road Map: Nanotechnology. (Courtesy of the CSJ). (see color insert)

Japan's Elements Strategy Initiative

In Figure 5 (CSJ Chemistry Dream Road Map for Inorganic Chemistry), several keywords deal with chemical elements. In this section, I will provide more details on Japan's Elements Strategy Initiative as a successful example of new concepts proposed by Japanese chemists. This was prompted by the decrease in the world's supply of elements, which is actually a more immediate problem than the decreased supply of oil. In 2004, Prof. Shinji Murai and I organized a two-day workshop at Hakone by gathering about 30 leading chemists to discuss materials science for the future. Through overnight discussions, more than 10 new concepts were proposed, including the Elements Strategy Initiative proposed by Prof Eiichi Nakamura.

The Japanese government considered our proposals seriously because Japan is a small island country without elemental resources and started to support several research programs starting in 2007 (Table 1). I have been serving as the supervisor of one program since 2010, covering 12 projects. The total budget of the initiative is roughly 40 B yen (400 M USD).

Table 1. "Elements Strategy Initiative" research projects in Japan

Year	Sponsor	Program
2007	MEXT	"Elements Science and Technology Project (ESTP)". 2007-2013: 16 projects
2007	METI	"Rare Metal Substitute Materials Development Project". 2008-2013: 10 projects for In, Dy, W, Pt, Ce, Tr, Eu
2010	CREST (JST)*	"Creation of Innovative Functions of Intelligent Materials, on the Basis of Elements Strategy". Supervisor: Kohei Tamao, 2010- 2017: 12 projects
2010	PRESTO (JST)	"New Materials Science and Elements Strategy". Supervisor Hideo Hosono, 2010-2017: 35 young chemists
2012	MEXT+ METI	"Elements Strategy Initiative Core Research Centers". (1) Permanent Magnet (T. Hirosawa, National Institute for Materials Science) (2) Catalysis and Batteries (T. Tanaka, Kyoto Univ.) (3) Electronics Materials (H. Hosono, Tokyo Institute of Technology) (4) Structural Materials (I. Tanaka, Kyoto Univ.)

* http://www.element.jst.go.jp/en/index.html.

This concept has caught on around the world. Thus, a trilateral EU-Japan-US conference on critical materials was held in 2010 in Washington, DC, followed by the second conference last year in Japan. These trilateral meetings had a political aim, putting pressures on China, which imposed export restrictions on some elements, especially rare earths.

More importantly, this concept greatly activated research in this area. As an example, US DOE announced the launch of the Energy Innovation Hub for critical materials research in May 2012 with a budget of 120M USD for 5 years. This hub is located at Ames Lab in Iowa and includes 18 universities and companies. One of the main purposes of this program is the development of the clean energy and the replacement of highly supply risk elements like dysprosium with elements of lower importance and lower risk.

This is our Message #2: *The Elements Strategy Initiative, the Critical Materials Strategy, and other related ones have now been recognized as global key strategies toward a sustainable society.*

CSJ's International Activities

CSJ has been involved with international activities. We have several types of activities in different areas and at different levels. Thus, we are involved with Asian Chemistry Congress (ACC) within Asia, the Chemical Sciences and Society Summit (CS3) for leading scientists in five countries, several bilateral and multilateral symposia for young chemists, and International Chemistry Olympiad for high school students. We are also interested in the Global Summit of Research Institute Leaders (RIL Summit).

CSJ is a member of CS3, together with ACS, RSC, CCS, GDCh. The CS3 is a closed workshop focusing on specific topics for a small number of leading scientists, about 40 in total (8 from each country), together with respective funding agencies (Table 2). The messages from the first CS3 have been particularly simple and impressive, viz.,

The science today is the technology of tomorrow.

Investing in chemistry is investing in the future.

The chemistry students today are the energy scientists of tomorrow.

As shown in the Table 2, the fifth meeting will be held in Japan in September 2013. All the meetings have resulted in white papers.

Table 2. Summary of CS3 Meetings

No.	Year	Country	Topic
1	2009	Germany	Sunlight to Power the World
2	2010	UK	Sustainable Materials
3	2011	China	Chemistry for Better Life
4	2012	USA	Next Generation Sustainable Electronics
5	2013	Japan	Efficient Utilization of Elements

We held several bilateral and multilateral symposia for young chemists, including those arranged with RSC, CCS, CSC, and the Asian International Symposium. With RSC, we signed an International Cooperation Agreement in July 2010. We had joint symposia for young chemists in 2007, 2008, 2010, and 2013. At the CSJ Annual Meeting at the end of March 2013, the PCCP Awards were presented by Dr. Robert Parker, Executive Director of RSC, to three young awardees.

With CCS, we signed an International Cooperation Agreement in Arpil 2009 and had Young Chemists Forums in 2010, 2012 and 2013. The joint symposium in 2013 focused on elements strategy.

With CSC, we had the first joint symposium during the CSJ Annual Meeting in 2013. On this occasion we invited Prof. Howard Alper as the honorary foreign member of the CSJ, the first from Canada. Cathy Crudden, President of CSC, also attended together with two young Canadian chemists.

Every year, we convene the Asian International Symposium by inviting about 20 young chemists from China, Hong Kong, India, Korea, Singapore, and Taiwan. All speakers are given distinguished lectureship awards.

Finally, although it was not CSJ's activity, I would like to make a brief mention of the First Global Summit of Research Institute Leaders (RIL Summit), which was held in Kyoto, Japan, in 2012. The purpose was to gather scientists from the world's leading institutes under one roof to discuss and share views on global scientific issues. The summit was intended to provide networking opportunities among the world's leading research institutions.

A summary of international activities from my perspective is given in Figure 9. The different types of activities can be organized as a layered structure, where the layers correspond to experience, from young scientists to leaders and senior statesmen. In this ACS symposium most of the speakers are organization leaders -- at level four. However, there is a paucity of global events for the next-generation scientists. So, I would like to propose the organization of more bilateral or multilateral symposia or workshops for the younger generations, e.g., graduate students, postdocs, and assistant professors.

At this point, I would like to propose Message No. 3. *Let us proceed with a stronger international cooperation towards development of young talent as the energy scientists of tomorrow.*

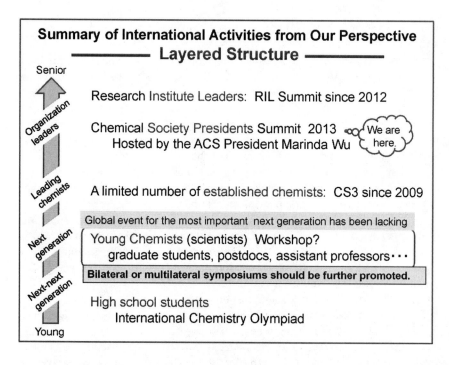

*Figure 9. Summary of international activities from the author's perspective.
(see color insert)*

Conclusions

In this article, I have provided an overview of the state of chemistry in Japan, particularly the impact of four main events in 2011. I have described Japan's Elements Strategy Initiative as an example of R&D directed towards national need. I have also reviewed many of the international activities that CSJ is currently involved in. My three main messages are summarized below:

1. In Japan we know we need to play a central role to overcome various problems [due to the Great East Japan Earthquake] and to restore the public's confidence in ST towards a sustainable society.
2. The Elements Strategy Initiative, the Critical Materials Strategy, and other related ones have now been recognized as global key strategies toward a sustainable society.
3. [For all our sister chemical societies in the world,] let us proceed with a stronger international cooperation towards development of young talent as the energy scientists of tomorrow.

Chapter 15

Challenges and Opportunities of the Chemistry Community in Taiwan After the Recent Economic Crisis

Chien-Hong Cheng[*]

President (2011-2012),
Chemical Society in Taipei Department of Chemistry,
National Tsing Hua University, Hsinchu, Taiwan, ROC-30013
[*]E-mail: chcheng@mx.nthu.edu.tw

What are the challenges and opportunities for the chemistry community in Taiwan? The oil price and the raw materials hit record high as the global climate is heating up. The situation is predicted to get worse as the time goes on. Unfortunately, we need to import every drop of the oil and all the raw materials from other countries to Taiwan; we simply do not have the resources except sunlight and water. How can we survive in a comfortable way under this situation? In this article, I will discuss our reactions in Taiwan to the global changes in terms of chemical education, chemical research, chemical industry and government policies. In addition, I will try to provide solutions to the challenges.

At the ACS Presidential Symposium ("Vision 2025: How to Succeed in the Global Chemistry Enterprise") in New Orleans, the author discussed the challenges and opportunities of the chemistry community in Taiwan after the recent global economic crisis.

An overview of the scope and strengths of the chemistry community in Taiwan may be necessary to understand the context of which the challenges and opportunities arise. Taiwan is host to the Chemical Society Located in Taipei, otherwise known as CSLT. The CSLT has approximately 2500 individual members and approximately 60 organizational members. There are 25 chemistry departments in universities across Taiwan at the time of the 2013 ACS meeting. There are also approximately 5 chemistry related institutions responsible for chemistry education and research.

The major universities and research institutions include at least National Taiwan University, National Tsing Hua University, National Chiao Tung University, National Cheng Kung University, Academia Sinica, Industrial Technology Research Institute, and many other public (e.g., NTNU, NCU, NCHU, NCCU, NSYU, NDHU and private institutions).

One potential for opportunity is the number of incoming chemistry students in Taiwan's universities, thus providing a powerful intellectual workforce. For example, in Table 1 the numbers of incoming chemistry students in Taiwan's universities are shown between the period of 2005 and 2011.

Table 1. The numbers of incoming chemistry students in Taiwan's universities between the period of 2005 and 2011[a]. (Source: the Chemical Society Located in Taipei)

Year	Undergraduate	Master	Doctorate
2005	1567 (1067/500)	899 (635/264)	206 (166/40)
2006	1740 (1128/612)	963 (683/280)	197 (155/42)
2007	1798 (1256/542)	966 (646/320)	208 (162/46)
2008	1753 (1145/608)	1003 (648/355)	184 (143/41)
2009	1725 (1086/639)	987 (639/348)	208 (151/57)
2010	1907 (1241/666)	1016 (679/337)	188 (142/46)
2011	**1715 (1101/614)**	**1040 (704/336)**	**153 (112/41)**

[a] The numbers are denoted as total (male/female students).

Comparatively, the number of graduating chemistry students in Taiwan's universities are shown in Table 2.

Table 2. The numbers of graduating chemistry students in Taiwan's universities are shown between the period of 2005 and 2010[a]. (Source: the Chemical Society Located in Taipei)

Year	Undergraduate	Master	Doctorate
2005	1393 (989/404)	740 (494/246)	136 (118/18)
2006	1364 (897/467)	811 (576/235)	124 (104/20)
2007	1475 (952/523)	800 (556/244)	149 (123/26)
2008	1502 (970/532)	853 (569/284)	130 (108/22)
2009	1718 (1119/599)	859 (562/297)	155 (128/27)
2010	**1587 (1070/517)**	**868 (567/301)**	**149 (107/42)**

[a] The numbers are denoted as total (male/female students).

The source of funding for basic chemistry research is generally provided by the National Science Council. There are approximately 500 principal investigators in chemistry institutions running approximately 600 projects. The total budget is 1.2 billion NT dollars per year, which is more or less equivalent to US $40 million per year. There are approximately 600 PhD students involved with the projects, approximately 1200 master students, approximately 66 assistants, and approximately 200 postdoctoral researchers.

In terms of quality of research, one potential estimator is the Scientific Citation index (SCI). Table 3 illustrates the ranking of SCI papers from Taiwan.

Table 3. The ranking of SCI papers from Taiwan based on Essential Science Indicators[a]. (Source: National Science Council, Taiwan)

Field	No. of SCI papers				No. of citations				Citations/paper			
	2009	2010	2011	2012	2009	2010	2011	2012	2009	2010	2011	2012
chem	15	15	15	15	18	18	18	17	33	33	32	32
phys	16	16	14	14	21	20	20	20	50	49	53	53
math	18	17	17	17	20	21	21	21	29	34	38	40
earth sci	25	24	23	21	24	24	24	25	51	52	49	52

[a] Updated as of January 1, 2013 to cover a 10-year + 10-month period.

Another crude way of measuring success of the chemistry enterprise is the chemical sales as compared to other countries. In Figure 1 Taiwan's sale of chemicals is compared to other top chemical exporters around the world.

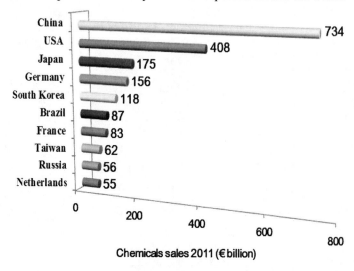

Figure 1. Chemical sales in 2011 in billions of Euros. (Source: Cefic, Facts and Figures, 2012).

After their education and research careers, the students in the chemistry departments often find jobs in the high-tech industry, in the chemical industry, in the chemical engineering industry, or the pharmaceutical industry. Figure 2 illustrates the total production value of Taiwan's chemical production in 2012.

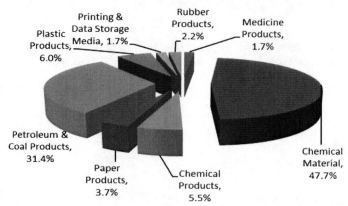

Figure 2. Total production value of chemical industry in Taiwan is US$144.4 billion (2012). The pie chart shows the sub-industries contributing to the total production value. (Source: Department of Statistics, Ministry of Economic Affairs and Industrial Technology Research Institute, Taiwan (USD:NTD = 30:1)).

Another indication of Taiwan's Chemical research talent is the total number of employees in the chemical industry. Figure 3 illustrates the employees of chemical industry in Taiwan.

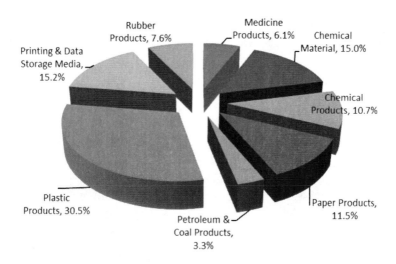

Figure 3. Total employees of chemical industry in Taiwan. The total employment is 452,806 (2012). (Source: Directorate General of Budget, Accounting and Statistics, Executive Yuan and Industrial Technology Research Institute, Taiwan).

As illustrated by the statistics in Figure 3, the opportunity for chemical industry is based on its talent pool and production. While Taiwan has very little raw materials, Taiwan does import raw materials and provide value added processing. However, this strategy is increasingly facing challenges from Taiwan's neighbors due partly to more cost-competitive research and development centers staffed by less costly human resources.

Taiwan has addressed this challenge by moving more to heavily value-added products, such as optoelectronic materials (e.g. LCD, LED, solar cell, OLED materials), medicinal products, printing and data storage materials, and special chemicals for the semiconductor industry. Figure 4 illustrates an example of how Taiwan has shifted their economical business model within the petrochemical industry. The three main future developments are: 1) maintain economic growth, 2) add no environmental burden, 3) promote energy efficiency. The strategies include low carbon emission, threngthened safety management, enhanced energy efficiency, and the development of high-value added product. Several companies and joint ventures in Taiwan are now involved with high value-added petrochemicals, such as ethylene glycol, C5 chemicals, and petroleum resins for production in 2013-2015.

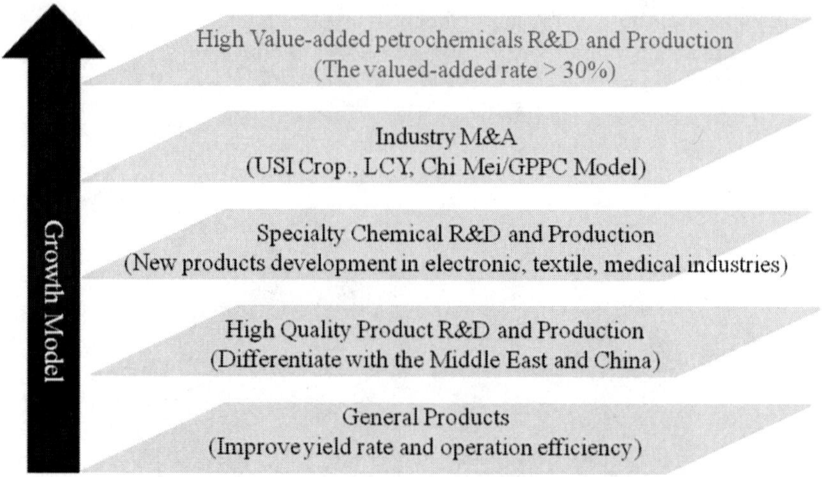

Figure 4. Taiwan's Petrochemical Industry Development Strategy. (Source: Industrial Technology Research Institute, Taiwan).

Conclusions

Despite the shortage of raw materials and energy, Taiwan has developed a capitalist economy that ranks as the 18th in the world by gross domestic product (GDP). The chemical industry accounts for about 29% of the GDP. However, Taiwan's chemical industry is facing many challenges today: competition from Asian neighbors, high oil prices, trade barriers, and international regulatory requirements. The Taiwan government has identified biotechnology, medical care, green energy, agriculture, optoelectronic materials and semiconductor among others, as strategic industries. Chemical industry is increasingly seeking to improve and upgrade its production technology and create value-added products. To achieve the goal, some chemical companies have sought collaboration with universities in research and development projects and in training and acquiring the the suitable researchers and engineers.

The CSLT is successfully adapting to these developments as it seeks to serve its members. Its annual meeting constantly attracts over 2000 participants with more than 1000 papers presented. It has organized many successful international chemical conferences, including Asian Chemical Congress (1999), IUPAC Macro (2008), and Pacific Polymer Conference (2013). In addition, the International Chemical Conference Taipei (ICCT) has been organized by CSLT since 1977. Recent themes included chemical education (2010), analytical chemistry (2008), materials and catalysis (2007), and synthetic chemistry (2005). The full scope of CSLT's educational, research, and publication activities is given in its website: www.chemistry.org.tw

Acknowledgments

We thank Industrial Technology Research Institute, National Science Council, Taiwan, the Chemical Society Located in Taipei and The EuropeanChemical Industry Council (cefic) for providing data for this article.

Chapter 16

Global Chemistry Research:
Where Is Africa on This Stage?

James Darkwa*

South African Chemical Institute, Johannesburg, South Africa
*E-mail: jdarkwa@uj.ac.za

The World Bank's forecast for Africa's economic growth for 2013 is 4.8%, at a time when most regions in the world have very low or negative growth of their economies. In fact, if one removes South Africa from this forecast, growth is expected to be 6% in Sub-Saharan Africa; yet Sub-Saharan Africa hardly features when it comes to Global Chemistry Research. A number of factors can be cited as contributing to this low level of chemistry research activities. Some of these factors are: lack of infrastructure, and lack of funding for research and graduate studies. There, however, have been a few success stories as a result of support received from the international community that has help improve funding for research and graduate training. These include programs by the International Science Programme (ISP) for sandwich programs for graduate training, and research grants from the International Science Foundation (IFS) and the Academy for the Developing World, formerly known as the Third World Academy of Science (TWAS). This article outlines the impact such supports have had on chemistry research on the African continent; but also suggests that these supports alone are insufficient in enabling Africa to play a significant role in global chemistry research unless the continent itself, finds other ways of funding the larger part of its chemistry research.

Introduction

"Science and technology" is seen as one of the things that will drive Africa's development, yet it is one area that has been neglected in the last two decades as Africa comes out its economic decline of the early 1980s. It is even more telling if one considers the fact that a number of African Universities, like University of Ibadan in Nigeria and Kwame Nkrumah University of Science and Technology (KNUST)) in the early 1970s had what were then considered state-of-the-art instruments. For instance, KNUST had a 60 MHz Nuclear Magnetic Resonance (NMR) spectrometers when even world class institutions in the early 1970s only had either a 60 MHz or a 100 MHz NMR spectrometers in their Chemistry Departments. Hence, one can say that KNUST was competitive then, but once this NMR spectrometer ceased to be functional in the 1980s, there has not been a functional NMR spectrometer in the whole of Ghana. One can therefore rejoice that the two oldest Universities in Ghana, namely University of Ghana and KNUST, through the initiatives of their Vice-Chancellors, have almost simultaneously recently placed orders for two high-field NMR spectrometers; a sign that institutions are beginning to invest in infrastructure for chemistry research. Similar investments in high-field NMR spectrometers have been made by institutions such as Addis Ababa University and University of Nairobi, but such purchases have only occurred in very few institutions.

Most countries also do not seem to have clear-cut policies for promoting research. So it is heartening to note that countries, such as South Africa have clearly defined research and development priorities that include aspects of chemistry research. It is therefore difficult to compare the chemistry research in Africa with most parts of the developed world. This article will thus only use key points to demonstrate where Africa is as far as global chemistry research is concerned, and in the process, draw attention to some of the interventions that have worked to promote chemistry research on the continent. In the end I hope this article will be able to demonstrate that outside support and interventions alone will not get chemistry research in Africa on the global stage unless African governments specifically make significant contribution towards chemistry research and science and technology research in general. I will begin by outlining where chemistry is in South Africa's research and development strategies.

Research Priorities in South Africa and Where Chemistry Fits in

Recently South Africa's Minister of Science and Technology announced that human and social dynamics, energy security, bio-economy, global change and space science as South Africa's priority research and development areas. While chemistry is not explicitly spelt out in these five priority areas, chemistry features prominently in energy security. The country has invested substantial amounts of resources in what the country terms "the hydrogen economy". Huge investments in biotechnology and nanotechnology have also been made, both of which have large contribution from chemistry.

National Initiatives with Chemistry-Related Research

About 6 years ago, the South African government made a conscious decision to embark on fuel cell research and development which is envisaged to add value to the country's vast deposits of platinum group metals. In order to pursue this objective, it set up three fuel cell research centers under the umbrella of Hydrogen South Africa, *HySA®*. The three centers are: *HySA/Catalysis, HySA/Systems* and *HySA/Infrastructure*. The catalysis center is charged to develop catalysts and catalytic devices that are used for hydrogen production, while the infrastructure center is to develop hydrogen production, storage and delivery devices and the systems center is to develop, build and commission prototypes. The annual funding for these three centers is estimated to be about $10 million annually. This is a clear indication of matching resources with research direction.

Another chemistry driven South African research initiative is the National Nanotechnology Innovation Centers (NICs). These are housed by the Council for Scientific and Industrial Research (CSIR) and Mintek, a research and development organization that specializes in mineral and metallurgical technologies. The NIC at the CSIR focuses on developing new materials using nanotechnology and their applications in manufacturing, energy and health; whilst the center at Mintek is focused on sensors, biolabels and water nanotechnology. Mintek's nanotechnology center appears to have benefited from an earlier project, project AuTek, which was commissioned to find new uses for gold. Under AuTek, Mintek developed the capacity to make gold nanoparticles for biomedical applications and this is what appears to have led to Mintek being chosen as one of the NICs.

In fact, metals play such an important role in the economy of South Africa that the Department of Science and Technology has another flagship project that center's around metals, namely the Advanced Metals Initiative (AMI). The AMI is also run via networks that are coordinated by three centers at the CSIR, Mintek and the Nuclear Energy Corporation of South Africa (NECSA). The AMI project at the CSIR coordinates a network development of light metals such as Mg, Al and Ti, while NECSA coordinates a network for the development of Zr, Hf, Nb and Ta, and Mintek coordinates a similar network for the development of Au and platinum group metals (PGMs). It is therefore clear that in South Africa much of chemistry-related research hinges on the country's mineral resources, giving it a competitive advantage when it comes to metals research globally; a lesson that the rest of the continent can learn.

Chemistry in Sub-Saharan Africa

With the exception of South Africa, research in the rest of sub-Saharan Africa has been generally hampered by a combination of the following factors: lack of clear-cut national policies and funding, inadequate infrastructure and lack of visionary leaders. As such, much of the funding for projects in sub-Saharan Africa has come from external sources, mainly Europe. Such support at times limits areas of research as prescribed by the donor, but most of the time research that can be conducted is limited by the available resources in a particular country. The above notwithstanding, there have been a number of excellent support programs that have allowed some research to be done in chemistry. Most of these programs have been made possible via support provided by various international organizations (Figure 1), mostly from European countries. I will describe a few of these programs in the subsequent sections of this article.

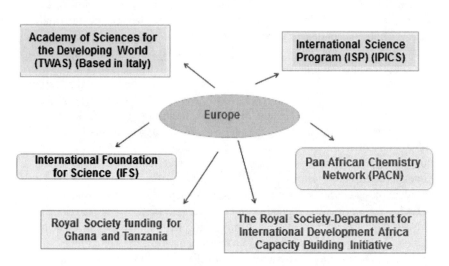

Figure 1. International organizations that support chemistry and other sciences in Africa.

The International Science Program (ISP)

The International Science Program (ISP), established in 1961, is based at Uppsala University in Sweden. It runs three sub-programs, namely: (i) the International Program in the Physical Sciences (IPPS, since 1961), (ii) the International Program in the Chemical Sciences (IPICS, since 1970) and (iii) the International Program in the Mathematical Sciences (IPMS, since 2002).

The ISP in general encourages regional collaboration in its model of support, but it also requires links with Swedish institutions that can provide expertise on the project as well as facilities that may not exist in the home institution. ISP collaborations are mostly driven via graduate student training, but also make room for the development of young faculty associated with a project. It also encourages the formation of networks that bring people and regional institutions together, and it is through these networks that I think IPS has made the most impact. Some of these ISP supported networks in Africa are highlighted later.

ISP also supports conferences that bring together its grantees. For example in 2009 ISP organized a conference on "Regional and Interregional Cooperation to Strengthen Basic Sciences Developing Countries" in Addis Ababa, Ethiopia (1); where most of the participants presented papers on activities that were either fully or partially supported by ISP. A survey conducted during this conference and reported in the proceedings, indicate that a number of African chemists complained of inadequate or no support for their projects by their own countries. The survey also showed how crucial networks are enabling chemists to work together. The work of some these networks reported below.

International Foundation for Science (IFS)

The International Foundation for Science (IFS) is another funding organization that has supported science research, including chemistry, in developing countries over the years. The IFS was established in 1972 and supports young researchers from developing countries; especially from Sub-Saharan Africa (Figure 2). I received my first research grant ($12 000) as a young researcher from the IFS in 1990 and this grant, to a very large extent, allowed me to start my independent research career. There are several other established African chemists who have benefited from IFS grants; so these grants have been invaluable, small as they may appear.

The IFS places a great deal of emphasis on research on agriculture, biological, water and energy resources; has natural products and organic as sub-theme. So it is therefore not surprising that most African IFS chemistry grantees work on natural products. This has led to two natural products networks on the continent which will be dealt with later. But it is also interesting to note that IFS support has gone to women (Figure 3). I must add, though, that the defined area of support for chemistry research by the IFS, lack of local support, has meant that other areas of organic chemistry research are less developed compared to natural products.

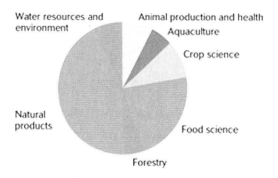

Figure 2. Distribution of IFS support for African scientists. (Source: IFS-OPCW survey of African scientists-2006). (2)

171

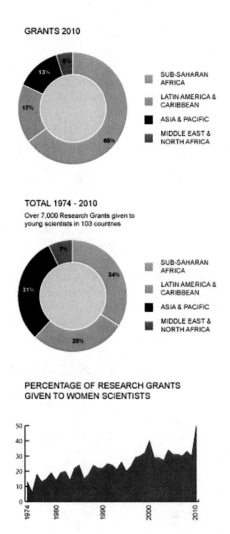

Figure 3. IFS grants from 1974-2010. (Figures taken from the IFS website).

The World Academy of Sciences (TWAS)

The World Academy Sciences (TWAS), formerly known as the Third World Academy of Sciences, is another organization that funds research for scientists in the developing world and thus by extension, African chemists. Like the IFS, it has funded chemistry research in Sub-Saharan Africa since the mid-1980s, and is another organization that funded my own research as a young chemist. The level of its funding is similar to the IFS, but the TWAS also funds "Centers" that consists of a group of researchers at a much higher level. Most young chemists could combine IFS and TWAS funding, like I did, to get a reasonable amount of grant for their research. The combined grant now comes to about $25 000, which can be a good source of chemicals and small equipment for any chemist on the African continent. This source of funding that could be further enhanced by support from the recipient countries because this combined support will nonetheless, be insufficient to run high level research projects.

Chemistry-Related Networks on the African Continent

The role networks have played and continue to play in the promotion of chemistry and chemistry research on the African continent cannot be over emphasized. Some of these networks have however, been more prominent than others. Three prominent networks for chemistry research are: (i) the Natural Products Research Network for Eastern and Central Africa (NAPRECA) (ii) The Network for Analytical and Bio-assay Services in Africa (NABSA) and (iii) the Southern and Eastern African Network of Analytical Chemists (SEANAC). I will thus describe activities of these three networks.

(i). NAPRECA

NAPRECA was established in 1984 to initiate, develop and promote research in the area of natural products chemistry in Eastern and Central Africa. It currently has 11 member countries that span from Sudan to Botswana. Its success has largely been owing to the fact that there is a variety of flora on the continent that provides large sources of chemicals for these chemists. I, however, attribute the success of this network, to a very large extent, to the leadership that this organization had in its early days. The leadership ensured that the organisation has good structures, financial support and eventually credibility. This leadership in the early days included Berhanu Abegaz and Ermias Dagne, from University of Addis Ababa. It is therefore not surprising that University of Addis Ababa continues to be seen as a Center of Excellence for African Natural Products Chemistry. The reader will see later the other roles that these two chemists have played and continue to play in natural products chemistry on the continent.

NAPRECA now holds regular conferences, workshops and training programs; but its main activity is graduate training and research. A good number of NAPRECA's graduate students are supported by fellowships provided by the German Academic Exchange Services (DAAD), a fellowship that has been in operation since 1988. In addition to fellowships from DAAD, NAPRECA's success has been underpinned by the services provided by NABSA, which provides analytical services for the analysis of natural products.

There is a West African counterpart to NAPRECA, namely the Western African Network of Natural Products Research Scientists (WANNPRES) (established in 2002), but WANNPRES has been much less successful than NAPRECA. The lack of success is largely due to a lack of effective leadership and resources.

(ii). NABSA

The formation of NABSA was necessitated by the dire need for analytical services, initially for natural products chemists, but eventually there was a need to provide analytical services to other chemists on the continent. Founded in 1992, again with the involvement of Berhanu Abegaz, NABSA was set up to promote the development of scientific activities of reasonable quality in Africa through sharing facilities and expertise. It has been largely successful due, to the excellent NMR and mass spectrometry facilities at University of Botswana (Figure 4). A NABSA report by Abegaz and Bezabih (3) highlights how this is an excellent example of Intra-African cooperation in chemical sciences. Key points in the report are: delivery of over 12 000 NMR and mass spectral data over a 10 year period and providing analytical data for the completion of 14 PhD dissertations by students from Tanzania and Cameroon. However, none of these achievements would have been possible without financial and other material support from IPICS (main funder), UNESCO, USAID, IFS, TWAS, Department of Science and Technology (DST) (South Africa), the Gates Foundation and University of Botswana. It is interesting to note the support from University of Botswana and DST, the latter's support will feature in other areas; demonstrating the quiet role South Africa is playing in funding science and technology research in Africa. The extent of NABSA's activities can be seen in Figure 4 and Table 1.

NAPRECA and WANNPRES have both benefitted from NABSA's support. It can therefore be said that NABSA is the fulcrum on which Africa's natural products research has revolved as demonstrated by Figure 5.

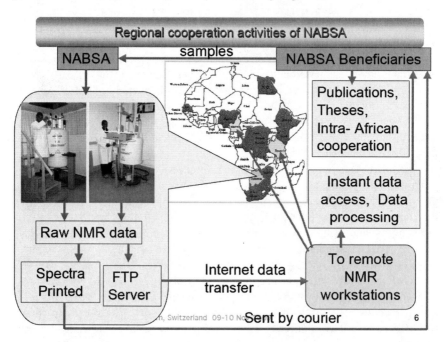

Figure 4. A summary of NABSA's activities. (4)

Table 1. Publications emanating from NABSA's activities. (*3*)

NABSA Publications – Intra-African

Journal	Ethiopia	Tanzania	Cameroon	Zimbabwe	Total
Biochemical Systematics and Ecology			3		3
Bulletin of Chemical Society of Ethiopia	2		4		6
Journal of Ethnopharmacology			3		3
Nat Prod Communications					1
Phytochemistry	6	2	13		21
Pure & App Chem			2		2
Others	6		14	2	22
Total	14	2	32	2	58

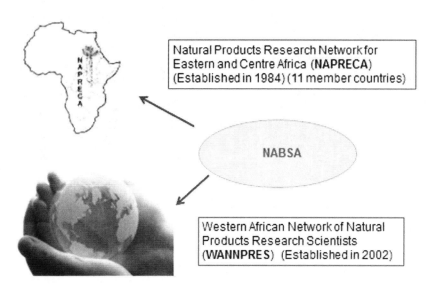

Figure 5. How NABSA supports NAPRECA and WANNPRES.

(iii). SEANAC

SEANAC is another successful network of chemists that brings analytical chemists from all over Africa together, although it was initially intended to be a network for East and Central African analytical chemists. Much of the success of SEANAC can be traced to the leadership provided by its founding Secretary General, Nelson Torto. SEANAC now has short term fellowships that allow young academics and graduate students to visit better equipped laboratories than their home laboratories in the network to use equipment and instruments that are now available in their home institutions. SENAC also holds conferences that allow members to present the results of their research to peers. Support for SEANAC, like the other networks, has come from various organizations one of which is the Organization for the Prevent of Chemical Weapons (OPCW).

Organization for the Prevent of Chemical Weapons (OPCW)

It is interesting to note that OPCW has become a great supporter of chemistry research in Africa. One can even go on to say that SEANAC has succeeded in its short term research visits primarily because of OPCW's support. OPCW has also in collaboration with IFS, provided support for workshops, training and scientific writing for African chemists. In 2006 there was a survey of African Scientists as part of a report on IFS-OPCW supported scientists about what these scientists considered their 12 difficulties in doing research on the continent. Figure 6 is a summary of the responses received. Their foremost difficulty, according to the survey, is financial support, followed by access to what can be interpreted as state-of-the-art equipment, consumables and laboratory space. These are all problems that can only be solved if African countries invest their own money to support research.

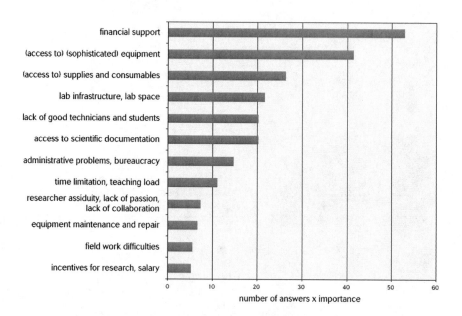

Figure 6. Weighted importance of 12 foremost difficulties for doing research. (Source: IFS-OPCW survey of African scientists-2006). (2)

International Organization for Chemical Sciences in Development (IOCD)

Another organization that has supported Chemistry in Africa is the International Organization for Chemical Sciences in Development (IOCD). Based in Belgium, the IOCD was founded in 1981 to promote the pursuit and application of chemical sciences for sustainable development in low and middle income countries. Its 2011-2020 three strategies (with their expected outcomes), for Africa and are listed below:

(a) **Chemistry for better health:** expected outcomes include capacity building for medicinal chemistry, including drug analysis, discovery and development (**Africa: African Network for Drugs and Diagnostics Innovation (ANDI)**)

(b) **Chemistry for better environment**: expected outcomes include enhanced capacity for environmental chemical analysis and sustainable use of biological resources.

(c) **Capacity building in chemical education:** expected outcomes include well-trained chemical scientists and suitable institutions and conditions in which they can work, including shipment of educational resources to Universities in Africa.

ANDI in particular is extremely important for the continent as it seeks to find end uses for natural products from the continent. This is by far the most extensive, and also the most sophisticated of all the networks (Figure 7). It also further indicates the level of involvement of South Africa in the continent's research.

South Africa has also supported science and technology research in Africa through bilateral agreements with a number of African countries. Some of these agreements are in Figure 8.

But Are These External Supports the Solutions to Funding Problems for African Chemistry Research?

It is clear from the above narration that some research in chemistry has been going on in sub-Saharan Africa, but the bulk of this research has been made possible largely by external support. The question that needs to be answered, however, is: "are these the solutions to funding problems for African chemistry research"? My answer to this question is No! Africa should start funding its own research and also drive its own research agenda. Fortunately it appears that the continent is beginning to act. Several academic institutions are beginning to build their own infrastructure for chemistry research; for instance KNUST and University of Ghana are buying a 400 MHz and a 500 MHz NMR spectrometers respectively; and this is very good news indeed. As stated earlier, these investments in scientific infrastructure by these institutions have been made possible because the Vice-Chancellors of the two Universities took it upon themselves to raise funds for these equipment. This is the calibre of leaders institutions need on the continent.

Other continental-wide organizations such as the African Academy of Sciences (AAS) are putting in structures to draw the attention of African governments to support science and technology research, which would include chemistry. The AAS has just released its strategic plan for 2013-2018. This plan, amongst other things, seeks to lobby the Africa Union and by extension African governments to honour their own pledge of investing 1% of their GDP in science and technology research. I would therefore request all the organizations that have supported research is Africa to work with the AAS in its mission to lobby African governments to fulfil this commitment as it is the only viable and sustainable way of funding research in general, and chemistry research in particular, on the African continent.

Figure 7. Organizations involved in ANDI.

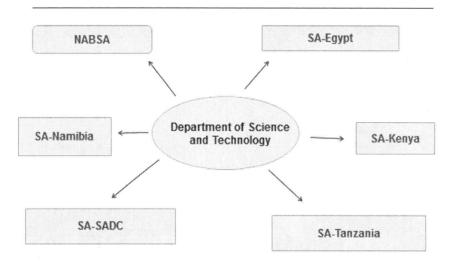

Figure 8. Some of the bilateral support for science and technology in Africa by South Africa.

Acknowledgments

This manuscript is based on the presentation made at the ACS Presidents symposium on Global Chemistry Research at the ACS meeting in New Orleans, April 7-11, 2013.

References

1. Kiselman C., Ed.; Regional and Interregional Cooperation to Strengthen Basic Sciences Developing Countries. In *Proceedings of an International Conference Organized by ISP*; Addis Ababa, Ethiopai, September 1–4, 2009.
2. Åkerblom, M. *International Foundation for Science*; MESIA Impact Studies Report No. 8; 2006.
3. Abegaz, B. M.; Bezabih, M. *Intra-African Cooperation in Chemical Sciences–NABSA's Efforts and Achievements.* http://www2.math.uu.se/~leifab/addis/posters/Abegaz-NABSA.pdf.
4. Abegaz, B. M. *Major Challenges for Research Institutes in Africa–A Mini-Survey of Opinions from former IFS Grantees*; IFS-KFPE Workshop, Bern, Switzerland, November 9–10, 2009.

Shared Experiences of Successful Global Start-Ups and Foreign Assignments

Chapter 17

Transforming Biopharma Innovation via Global Collaboration

Tao Guo*

Department of Medicinal Chemistry, WuXi AppTec Co., Ltd., Shanghai 200131, China
*E-mail: guo_tao@wuxiapptec.com

Global collaboration is transforming biopharma innovation. The biopharma industry is facing tremendous challenges from an increasing number of patent-expiring existing drugs to an increasing cost to develop innovative new drugs. In order to improve productivity and reduce cost, global biopharma has been increasingly outsourcing its R&D services. WuXi AppTec, founded in 2000, has built a comprehensive integrated drug discovery platform with the vision to improve the success of research and shorten the time of development. Having grown from four founders to over 7,000 employees, including over 3,000 chemists, with operations in both China and the US, WuXi AppTec has become the valued collaboration partner of choice for global biopharma. With its integrated drug R&D platform, WuXi is transforming innovation for global biopharma to allow any company use of the platform to discover and bring to market innovative medicines addressing unmet medical needs more quickly and cost-effectively. This paper will examine what it takes to be successful in the globalization of today and the future and how to go about seeking and seizing the opportunity.

Introduction

The biopharma industry, including small and large pharmaceutical and biotechnology companies, experienced a decline in the forecasted growth, based on prescription drug sales globally. "Prescription drug sales forecast falls $26bn to $827bn in 2016 vs. forecasts in May 2011 (*1*)"

Reasons for the decrease are attributed to the so-called patent cliff and a general decrease in R&D productivity based on the traditional biopharma business model that was successful up until the early 2000's.

2012 was the most significant time frame for companies facing a patent cliff, the top ten pharmaceutical companies recognized a collective loss of $33 billion of prescription drugs sales due to introduction of generic drugs post patent while forecasts for the next 6 years show a potential loss of over "$290 billion of sales at risk from patent expirations in 2012-18 (*1*)"

The loss of sales from approved patented drugs only begets the question as to the overall productivity of the industry. How are companies approaching filling their pipelines to replace revenue lost as patents on existing drugs expire? Will early stage and discovery level research be neglected in pursuit of late stage candidates? Will companies decrease investment in R&D to show respectable margins while facing lower revenues?

WuXi feels that there have been fewer 'shots on goal' coming from the traditional R&D model, and has invested in building a platform of R&D capabilities that any company or anyone can access to reach their goals faster and at lower costs.

Patent Cliff

"For the pharmaceutical industry, the last days of 2012 will mark the end of the patent cliff, an approximately 18-month stretch during which major drug companies lost exclusive rights to many billion-dollar-selling drugs (*2*)."

As an example of the devastating effects of the patent cliff we can look at Merck's top selling drug Singulair, a leading asthma and allergy drug. In 2011, this single therapy accounted for U.S. sales of $3.5 billion and global sales of $5 billion. Within four weeks after the patent expiration, Aug. 2012 in the U.S. and Feb. 2013 in major European markets, generic copies of the drug (Leukotriene D4 antagonist) were introduced and Merck faced a 90% drop in sales revenue (*1*).

"Market-watchers say the generic competitors will quickly take two-thirds of the market share for the drug's sales, with Datamonitor going farther to say generic versions will take over 90 percent of the sales in the United States within a year (*3*)."

Decreasing R&D Productivity

"The number of new drugs approved by the US FDA per billion USD (inflation-adjusted) spent on R&D has halved roughly every 9 years (*4*)."

While the patent cliff loomed menacingly over the prospects of the industry's top biopharma companies, publically held companies worried about showing a profit with plateauing or diminishing revenues.

One strategy was to reduce investment in discovery level R&D and invest in later stage prospects. Investment in R&D spending on preclinical candidates or compounds at later stages has remained high, and even increased. However, the increase, in the seven years from 2011-2018 of an average 1.5% compound annual growth rate (CAGR), was much lower than the average increase for the seven years from 2004-2010 of 6.5% CAGR (*1*).

Deloitte and Thomson Reuters conducted an R&D productivity study in which 2012 study results showed "the number of new drug approvals increased by around 30 percent, yet the expected revenue from these medicines actually fell by a similar amount (*5*)."

How are companies planning to fill the pipeline in the coming decade while facing significant decrease in revenues?

While cutting discovery costs is a short term solution to showing profits, a long term strategy needs to be developed. In the last few years we have witnessed a plethora of new business model ideas come to the fore. Purchasing preclinical candidates from biotechs, alliances with academic institutes, strategic outsourcing partnerships, and even partnerships or consortiums among competitors have made headlines. So far, the first strategy of buying into a viable entity has shown the most promise. Among the other three strategies, academic and consortium strategies may take ten years to be truly evaluated. Strategic outsourcing, with the right partner is most likely to show results in the next five years.

Outsourcing discovery R&D is a recent innovation, as most top biopharma companies considered targets, target validation, and mechanism of action know-how to be a core competency and the ultimate secrete that separated them from the competition increasing their chance of reaching the market first.

WuXi has seen in the last five years an increase in discovery R&D outsourcing. Starting from our foundation as a chemistry powerhouse, we recognized a real need from our partners for a full medicinal chemistry approach to save time and cut costs when developing leads. This expanded into lead optimization and eventually into the preclinical work conducted in cooperation and with the know-how of the lead development teams. The trust extended in this increased willingness to outsource can be seen in today's novel partnerships. Only five years ago, competitors would not have joined forces nor would top biopharma have handed their target ideas to academic institutions.

The loss of the blockbuster 'one size fits all' approach with the simultaneous rise in the new paradigm of personalized medicine also accounts for loss of R&D productivity. Recognizing the change in the science and market segments and investing in new technology and training can be demanding for long standing R&D teams.

"It has been eye-opening to realize how significantly we will need to invest in transformative partnerships and the broader health care ecosystem to be relevant in 3.0," says Kim Park of Johnson & Johnson's Janssen Healthcare Innovation Unit (*6*).

As mentioned, cutting costs is an easy start, but increasing productivity is a larger solution. "With an average internal rate of return (IRR) from R&D in 2012 of 7.2 percent -- against 7.7 percent and 10.5 percent in the two preceding years – drug makers are barely covering their average cost of capital, estimated at around 7 percent (7)."

A key to WuXi's success with discovery outsourcing goes back to the 'shots on goal.' As the low hanging fruit are harvested and new targets become harder to qualify, the traditional screening approach has recognized a higher attrition rate than in the preceding ten years. Now, a high throughput approach is still productive, but requires more sophisticated techniques to simultaneously look at mechanisms of action.

The WuXi advantage in this scenario is the sheer number of scientists and broad spectrum of techniques that can be applied to a target (Figure 1). WuXi has a large staff of highly skilled scientist, that partner with clients offering a flexible and eminently expandable capacity. This increase brings a real increase in throughput and a greater number of viable candidates, yielding more preclinical candidates in less time.

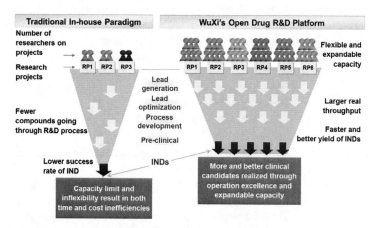

Figure 1. WuXi's open drug R&D platform helps partners to improve success and shorten development. Larger capacity and more data yield better compounds, thereby increasing success rate of discovery and shortening the time to market, which will result in tremendous savings for the industry and companies.

Core Strength: Our People

According to a recent study, scientists at outsourcing firms have "Contributed to the development of all of the top 20 selling prescription medicines and were involved in the development of at least 33 of 38 new medicines approved for use last year in the United States and Europe (8)"

"At a time when many people are asking "where are the jobs?" we have a clear answer with the tremendous growth that contract research organizations (CRO) have experienced over the past 10 years," said Doug Peddicord, PhD, Executive Director of Association of Clinical Research Organizations (ACRO).

"Our members continue to expand the range and scope of their services, from drug discovery through post-approval studies, developing deep expertise in complex areas like vaccines, biosimilars, comparative effectiveness research, and regenerative medicine (8)."

WuXi's core strength is our people (Figure 2). WuXi AppTec today employs more than 7,000 people globally. WuXi has focused on hiring seasoned leaders from top academic and industry institutions. Our theory is that bringing the brightest minds together is similar to the consortium approach being explored by top biopharma today. Our leadership teams have an average of fifteen, or more, years' experience and a proven ability to bring discovery concepts to the patients. From authoring patents for their institutions to publishing articles in scientific journals, our leaders are at the top of their field. This seasoned leadership, 70% of which have a MS or higher, is the key to ensuring that the new graduates we hire every year have the mentorship to reach their potential.

With thousands employees and customers, stellar program management has become the essential to our shared success. WuXi recognized the need for experienced program managers since the early days of large library synthesis. Now, our program management teams include over 300 western trained professionals with 5-10 years of industry experience. The program management program has been a resounding success, managing over 80 programs across all major therapeutic areas. Over 12 preclinical candidates were delivered in 2012.

Figure 2. WuXi's core strength is our people.

WuXi Vision

WuXi has been building a bench to patient platform for over ten years. Our idea is to make our technology and our know-how accessible for any size company to accelerate innovation, lower the costs and shorten time to market. Like our customers who test new ideas, WuXi has a four tiered approach to making this open platform successful. We endeavor to constantly enhance our capabilities, work closely with customers to transform innovation, broaden our reach to meet global needs, and fuel innovation by investing in promising companies.

'Enhancing our capabilities' means investing in new technologies and partnering with resources that marry well with our existing technologies. Lowering costs of R&D has been an ongoing strategy for many of our customers, but still the cost of developing an asset has risen since 2010, "The average cost of developing an asset between 2010 and 2012 has increased by four percent, from $1,089 million in 2010, to $1,137 million in 2012." In addition, the cost of failure for discovery compound increased 6% per asset (5). WuXi's approach has been to focus on investing in the discovery or 'front end' of innovation in order to help our clients build pipelines. Recently we have invested in later stage technologies to keep up the momentum to the clinic and market.

One result of this investment has been a transformation in innovation. Partnerships with our large pharma customer have led the way to more strategic outsourcing paradigms. Our partners recognize our commitment and benefit from our investment to help reach their pipeline goals at a margin that pleases their stakeholders.

More recently, WuXi began an investment fund to kick-start promising companies. Starting with Hua Medicine in 2011, to Novira and Foundation Medicine in 2012, we believe these innovators are on the verge of breakthrough technologies and therapies.

Our latest approach is to help launch novel therapies in China, which is the fastest growing healthcare market in the world, and eventually can be a platform for global launch.

Global Collaborations

WuXi PharmaTech joined forces with AppTec Laboratories in 2008 with the aspiration to providing global solutions to the biopharma industry. Today, the combined company WuXi AppTec has over 12 total R&D sites, with offices in Europe as well.

Five years ago, the advantage of global collaboration was access to high quality low-cost scientific skills in China combined with expert scientific, regulatory and market access experience in the US and EU. Today the advantage of global collaboration is broader, still taking advantage of scientific know-how in both China and the US but now looking towards China's fast growing healthcare market as a significant unmet need.

WuXi AppTec is well positioned globally to provide efficient, cost-effective value across the entire spectrum of research, preclinical development, manufacturing, and commercialization.

In Conclusion

The biopharma industry is at a cross roads, but the horizon looks promising. Outsourcing, once exclusively in the clinical field, or isolated to niche technologies in R&D fields, has become one of the ways forward through the 'storm'. WuXi has had the advantage of moving quickly to offer a trusted, efficient, and low cost solution for R&D, growing from early discovery all the way to the clinic. Combining resources in China and the U.S. was another approach to bring together global competencies to accelerate development. Our four tiered strategy supports innovation and transformation. For chemists looking for jobs today, a well-respected global outsourcing provider is an optimal choice.

Acknowledgments

Thanks are due to Megan Rooney at WuXi AppTec for her help on the manuscript preparation.

References

1. World Preview 2018 – Embracing the Patent Cliff, 2012. EvaluatePharma. http://www.evaluatepharma.com/worldpreview2018.aspx.
2. Mullin, R. *Chem. Eng. News* **2012**, December 10, 15–20.
3. Bouley, J. FDA Approves Generic Versions of Singulair, August 7, 2012. http://www.drugdiscoverynews.com/index.php?newsarticle=6427.
4. Scannell, J. W.; Blanckley, A.; Boldon, H.; Warrington, B. *Nat. Rev. Drug Discovery* **2012**, *11*, 191–200.
5. Measuring the Return from Pharmaceutical Innovation 2012: Is R&D Earning Its Investment?, 2012. http://www.deloitte.com/assets/ Dcom-Switzerland/Local%20Assets/Documents/EN/LSHC/ ch_en_Measuring_the_return_from_pharmaceutical_innovation_2012.pdf.
6. Progressions: Building Pharma 3.0, 2011. Ernest and Young (EY). http://www.ey.com/GL/en/Industries/Life-Sciences/Progressions--building-Pharma-3-0.
7. Zhang, M. Big Pharma Facing Patent Cliff and Declining R&D Returns, December 5, 2012. http://www.ibtimes.com/big-pharma-facing-patent-cliff-declining-rd-returns-919311.
8. ACRO Profile, October 3, 2011. http://www.acrohealth.org.

Chapter 18

Building Research Businesses on Integration of Basic and Applied Research: Value Creation and New Opportunities for the Chemical Enterprise

Mukund S. Chorghade*

Chorghade Enterprises, 14 Carlson Circle,
Natick, Massachusetts 01760, United States
*E-mail: chorghade@gmail.com

The chemical enterprise has undergone unprecedented changes due to mergers, acquisitions, loss of patent protection: a paucity of new products has created an "innovation deficit". Rapidly increasing pace of regulatory reform allied with the necessity of effecting drastic cost-reductions have resulted in strategic paradigm shifts. The signing of the GATT accords has also paved the way for collaborations in numerous areas of science. We will explore how we obtained mutual benefits by eliminating current challenges via sophisticated technology, strategic collaboration, global commerce and refined logistics. We have operated several businesses in the USA and abroad. Innovation is global and scientists should not hesitate to accept overseas assignments. It is important to adjust to certain cultural parameters but the adventure is well worth the risk. New opportunities exist for chemists in business development, regulatory affairs, project management, and strategic alliances.

Pharmaceuticals – Challenges and Opportunities

The pharmaceutical sector has traditionally been a vibrant, innovation-driven and highly successful component of industry at large. In recent years, a confluence of spectacular advances in chemistry, molecular biology, genomic and chemical technology and the cognate fields of spectroscopy, chromatography and crystallography have led to the discovery and development of numerous novel therapeutic agents for the treatment of a wide spectrum of diseases. In order to facilitate this process, there has been a significant and noticeable effort aimed at improving the integration of discovery technologies, chemical outsourcing for route selection / delivery of active pharmaceutical ingredients, drug product formulations, clinical trials and refined deployment of information technologies. Multi-disciplinary and multi-functional teams focusing on lead generation and optimization have replaced the traditional, specialized research groups. To develop a drug from conception to commercialization, the biotechnology / biopharmaceutical industry (which has been highly entrepreneurial) has reached out and established global strategic partnerships with numerous companies.

The pharmaceutical industry primarily in the U.S.A. and Europe has undergone unprecedented changes in recent years due to mergers and acquisitions. The rapidly increasing pace of regulatory reform allied with the necessity of effecting drastic reductions in the price of bulk drugs have also resulted in marked shifts in the strategic paradigms in this industry. Numerous corporations are seeking strategic partnerships overseas to enhance their global capabilities for drug discovery and development. Pre-requisites like a highly trained and motivated work force, political stability, and the formidable process research skills of the chemists make for a winning combination. The recent signing of the GATT and WTO accords has also paved the way for collaboration in the area of new drugs, biotechnology and agricultural products. As expected, significant strategic outsourcing is on the increase, particularly to India and China.

India is the home of one of the world's oldest civilizations and several of the greatest religions. It arouses a mixture of amazement, awe and fascination among many foreigners. It occupies the seventh largest land mass in the world; nearly a billion people living in its boundaries make it the second most populous country. India is a melting pot of a myriad of races, languages and religions; has the most striking disparities in wealth and poverty; and operates with primitive as well as sophisticated technologies. These dichotomies in the world's largest democracy have lent an exotic flavor to India.

The country was ruled for nearly a millennium by a disparate succession of local and colonial rulers; the economy was largely feudal and agrarian. The growth of science and technology in India in the post-independence (1947-) era has been impressive. The country possesses a well-developed Western style legal/administrative structure, a proficient and competent civil service administration and considerable support and enthusiasm among the populace for betterment of life through science. The academic and publicly funded institutions have provided a steady stream of multi-lingual, mainly English speaking, educated manpower. The scientific manpower pool is the second largest in the world; many of the country's technocrats, scientists, and engineers have been trained in the finest laboratories of Europe and the USA. Several researchers have won international awards and have published and lectured abroad. Chemical research has increased in breadth, sophistication, and finesse. Modern instrumentation is readily available; numerous laboratories, pilot plants, and manufacturing facilities conforming to stringent specifications of GLP and cGMP have been established. These facilities have received regulatory approvals from international bodies such as the US FDA, MCA-UK, TGA-Australia, MCC-South Africa and the WHO. Total Quality Management programs are in place in the leading professionally managed institutions, with noteworthy improvements in product quality and reliability. The industrial base is therefore strong and technologically sophisticated. The Indian Government's economic liberalization program has resulted in abandonment of the stifling protectionism of "Faubian and Nehruvian socialism." Trade barriers have been lowered, taxes cut, and bottlenecks for foreign investment have been removed. The Government of India has relaxed drug price controls and provided fiscal incentives to promote collaboration with the Western World.

The Pharmaceutical Sector in India

India is emerging as one of the largest and cheapest producers of therapeutics in the world, accounting in volume for nearly 8.5% of the world's drug requirements. The Indian pharmaceutical sector has achieved global recognition as a low cost producer of bulk chemicals and formulation products.

In the initial years of independence, the industry was monopolized by a few multinationals. A decade later, the industry showed signs of doing away with multinational dominance with the emergence of Indian companies with capacity for production of formulations based on imported bulk drugs. The seventies saw the emergence of bulk drugs manufacturing based on imported as well as indigenous technologies. In the eighties, the Indian R&D contributions became significant and imports of bulk drug technologies reduced drastically. The industry today manufactures practically the entire range of therapeutic groups; is nearly self-sufficient in raw materials, and its level of operation is on par with international standards in production, technology and quality.

The R&D strengths encompass world class expertise for organic synthesis and facilities for isolation and structure elucidation, biological screening. The essential complement of expertise in chemistry and infrastructure facilities allied with strong institutional linkages built up with various universities and pharmaceutical industry ensures successful up scaling, seamless technology transfer and implementation of technology.

The avenues of cooperation that have been exploited by various multinational pharmaceutical, biopharmaceutical companies are listed below as illustrative examples of the enormous benefits that could accrue worldwide.

1.   Synthesis of analogs for broad spectrum and high throughput screening
2.   Lead optimization and analog design

3. Designed organic synthesis, scaffolds and building blocks for lead generation and development of synthetic methodologies
4. Route selection, Process chemistry: preparation of 1-5 Kg. of drug candidates for pre-clinical and Phase I evaluation
5. Technology development and transfer to contract manufacturing of sunset molecules at the end of the patent production period
6. Strategic licensing of compounds discovered in India: Several groups have advanced programs in the areas of anti-infective, anti-histamine, CNS drugs, cardiovascular, and natural products based drug discovery
7. The pharmaceutical industry has seen several successes in process chemistry driven innovation. Pioneering research in carbohydrate chemistry led to the novel rearrangement (Figure 1). Successful scale-up led to the commercial indigenous and overseas production and manufacturing of the popular anti-HIV drug AZT (*1*) at a much lower cost. Further work resulted in cost effective processes for Stavudine and Lamuvudine.

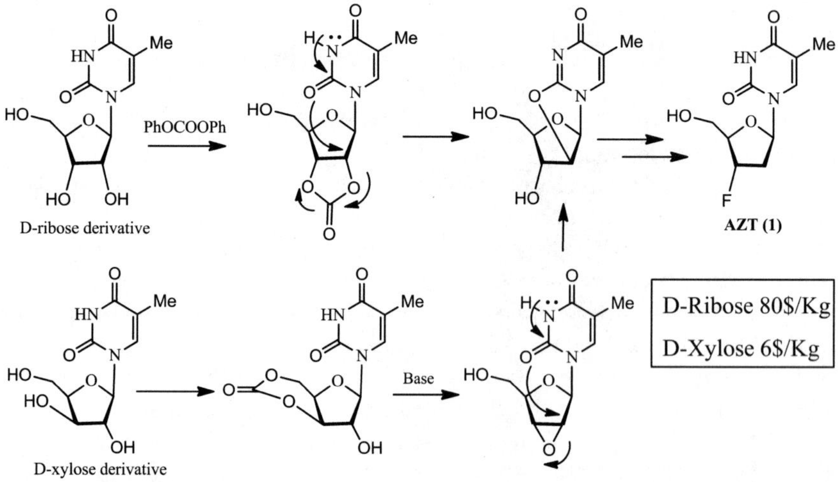

Figure 1. Reaction pathway leading to two drug molecules.

While biotechnological advances, genomics and high throughput screenings or combinatorial and asymmetric syntheses have opened new vistas in drug discovery, the industry is facing a serious innovation deficit. Critics suggest that "we have become high throughput in technology, yet have remained low throughput in thinking". Post marketing failures of blockbuster drugs have become major concerns of industries, leading to a significant shift in favor of single to multi targeted drugs and affording greater respect to traditional knowledge. Typical reductionist approach of modern science is being revisited over the background of systems biology and holistic approaches of traditional practices. Scientifically validated and technologically standardized botanical products may be explored on a fast track using innovative approaches like reverse pharmacology (Figure 2) and systems biology, which are based on traditional medicine knowledge (*2*).

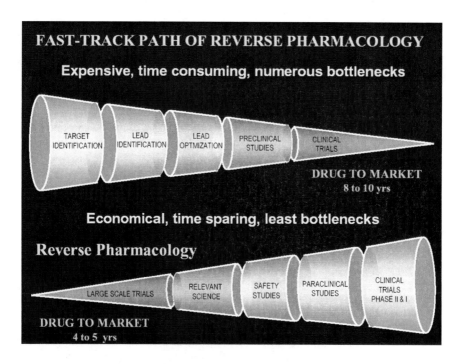

Figure 2. Conventional pharmacology (top) and reverse pharmacology (bottom).

Traditional medicine constitutes an evolutionary process as communities and individuals continue to discover practices transforming techniques. Many modern drugs have their origin in ethnopharmacology and traditional medicine. Traditions are dynamic and not static entities of unchanging knowledge. Discovering reliable 'living tradition' remains a major challenge in traditional medicine. In many parts 'little traditions' of indigenous systems of medicine are disappearing, yet their role in bioprospecting medicines or poisons remains of pivotal importance. Indian Ayurvedic and traditional Chinese systems are living 'great traditions'. Ayurvedic knowledge and experiential database can provide new functional leads to reduce time, money and toxicity - the three main hurdles in the drug development.

We begin the search based on Ayurvedic medicine research, clinical experiences, observations or available data on actual use in patients as a starting point. We use principles of systems biology where holistic yet rational analysis is done to address multiple therapeutic requirements. Since safety of the materials is already established from traditional use track record, we undertake pharmaceutical development, safety validation and pharmacodynamic studies in parallel to controlled clinical studies. Thus, drug discovery based on Ayurveda follows a 'Reverse Pharmacology' path from Clinics to Laboratories (2).

A Case Study: CMI-977

CMI-977 is a potent 5-lipoxygenase inhibitor that intervenes in the production of leukotrienes and is presently being developed for the treatment of chronic asthma. It is a single enantiomer with an all-trans (2S,5S) configuration. Of the four isomers of CMI-977, the S,S isomer was found to have the best biological activity and was selected for further development. The enantiomerically pure product was synthesized on a 2-kg scale from (S)-(+)-hydroxymethyl-γ-butyrolactone (*3*).

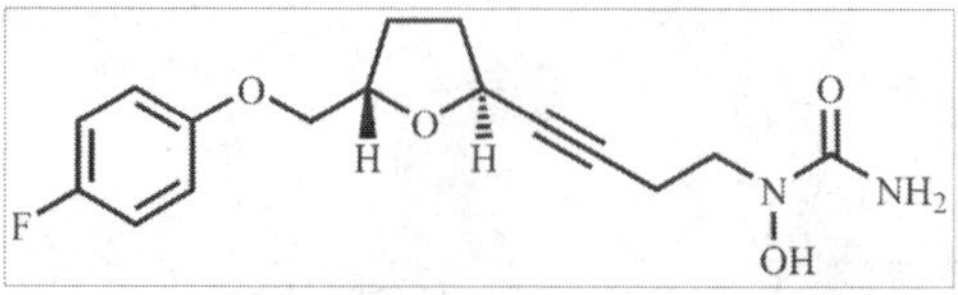

(2S,5S)-Trans-5-(4-fluorophenoxymethyl)-
2-(1-N-hydroxyureidyl-3-butyne-4-yl)tetrahydrofuran

In a later publication (*4*), an enantiomerically pure product was synthesized on a 2-kg scale from (S)-(+)-hydroxymethyl-γ-butyrolactone. A practical gram scale asymmetric synthesis of CMI-977 is shown in Figure 3. A tandem double elimination of an α-chlorooxirane and concomitant intramolecular nucleophilic substitution was used as the key step. Jacobsen hydrolytic kinetic resolution and Sharpless asymmetric epoxidation protocols were applied for the execution of the synthesis of the key chiral building block.

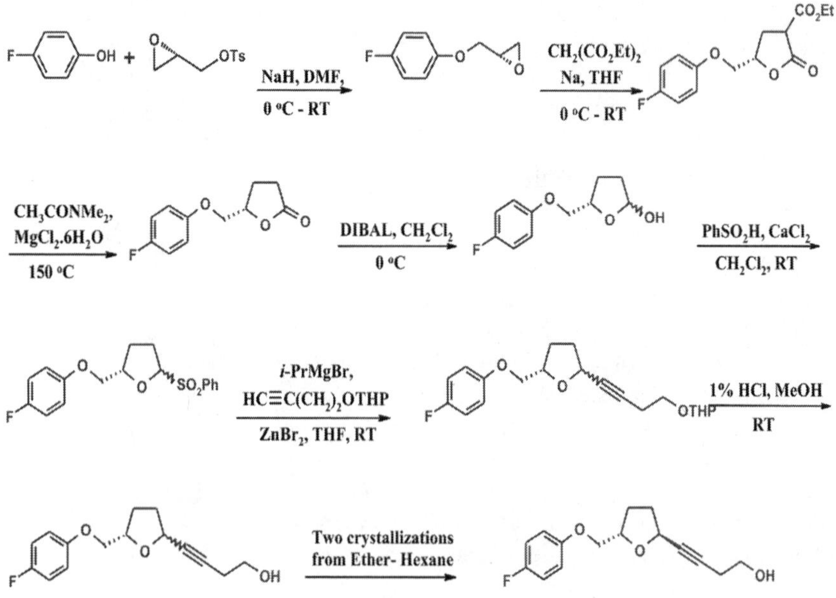

Figure 3. Novel synthesis of CMI-977 (ref. (4)).

Many more examples of drug discovery and development can be found in a book edited by the present author (5). It is useful to combine creativity with sophisticated technology, strategic collaboration, global commerce and refined logistics to produce new drugs and new businesses. The author has successfully operated several businesses in the USA and abroad. Innovation is global and scientists should not hesitate to accept overseas assignments. It is important to adjust to certain cultural parameters but the adventure is well worth the risk.

References

1. Rao, A. V. R.; Gurjar, M. K.; Lalitha, S. V. S. *J. Chem. Soc., Chem. Commun.* **1994**, 1255 (U.S. Patent 5596087, 1997).
2. Patwardhan, B.; Vaidya, A. D. B.; Chorghade, M.; Joshi, S. P. *Curr. Bioact. Compd.* **2008**, *4*, 1–12.
3. Cai, X.; Chorghade, M. S.; Fura, A.; Grewal, G. S.; Jauregui, K. A.; Lounsbury, H. A.; Scannell, R. T.; Yeh, C. G.; Young, M. A.; Yu, S.; Guo, L.; Moriarty, R. M.; Penmasta, R.; Rao, M. S.; Singhal, R. K.; Song, Z.; Staszewski, J. P.; Tuladhar, S. M.; Yang, S. *Org. Proc. Res. Dev.* **1999**, *3* (1), 73–76.
4. Gurjar, M. K.; Murugaiah, A. M. S.; Radhakrishna, P.; Ramana, C. V.; Chorghade, M. S. *Tetrahedron: Asymmetry* **2003**, *14*, 1363–1370.
5. Chorghade, M. S., Ed.; *Drug Discovery and Development*, Vol. 1; Wiley, Hoboken, NJ, 2006.

Chapter 19

New Paradigms for Educating Chemistry Professionals in a Globalized World

Hai-Lung Dai*

Temple University, Philadelphia, Pennsylvania 19122, United States
*E-mail: hldai@temple.edu

In today's ever more collaborative environment for education and research in a globalized world, what lessons can we take from other societies and what new avenues can we explore in educating future generations of chemists, or for that matter, scientists? The possibilities range from new paradigms for educating secondary science teachers, to collaborative dual degree programs that enhance students' professional capabilities, to dual doctoral degrees that enrich both the student's experience as well as the participating institutions' ability to conduct research. Teacher education programs, dual bachelor-master degrees program, and bilateral doctoral degree programs established at the University of Pennsylvania and Temple University will be highlighted as examples.

In this first century of the new millennium, technology continues to be the dominating force shaping up the economic landscape of the world. Economists have shown that since WWII, new technology and population growth account for nearly all economic growth in the US. Today, as many examples show, technology continues to drive changes in human practices and economic development. And the demand for new or better technologies has never been more important in addressing pressing societal needs. Among the most influential new forces along with technology is globalization, brought on by ever more efficient means for communication and transportation, and by the liberalization of national boundaries. Globalization on one hand means larger markets around the world for American goods and services, but it also brings more competition for human resources and capitals for the US. The ongoing globalization of research, development and business, coupled with the rapid economic development of countries around the world are heightening economic competition for US industries. The changes in the economic landscape unavoidably pose at the same time new opportunities as well as new challenges for higher education institutions in the US.

The 2005 National Academies' whitepaper "Rising above the Gathering Storm: Energizing and Employing America for a Brighter Economic Future" detailed the challenges the US is facing in science and technology in a globalized world, and laid out strategies, including those for higher institutions of learning, that the US should adopt to overcome these challenges. In education, the report detailed specifically the lack of preparation in math and science of American children in precollege education and the generally low interests by American youngsters in pursuing STEM related careers.

One of the major factors contributing to the lack of preparation for and interests in STEM careers by American youngsters is the shortage of content-prepared teachers in K-12 education. In response to the National Academies' whitepaper, which presented four recommendations with number one calling for more math and science teachers, President George Bush in his 2006 State of the Union address called for dramatic increase in numbers of math and science teachers to staff the K-12 classrooms.

The shortage of content-prepared math and science teachers is one problem American higher education institutions are partially responsible and can help to resolve. In most of the world outside of the US, teachers are trained through the so called 'normal' universities – institutions that are similar to comprehensive research universities here but have the unique mission of training public servants including teachers. A chemistry student in the National Taiwan Normal University or Zhejiang Normal University, for example, would take the full chemistry bachelor degree courses while also learn about teaching pedagogy. If a NTNU or ZJNU student decides to become a teacher, this student would take on additional classroom experience through internship to meet the teaching certification requirements. In the US, such training is going through the Education colleges within a university. A student in an education college is geared to take an education major focusing on pedagogy. In preparation to be a math or science teacher, the student often takes a number of math/science courses. Some decides on the number of courses based on state certification requirements, which vary from state to state. In some states, certification only requires as few as 10 credit hours of courses. The consequence of such teacher preparation practice is that the teacher may know a lot about education theory and have ample teaching experiences, but often does not have enough content in the subject area to teach students.

To overcome this deficiency in content preparation in math and science teachers, there have been a few innovative initiatives in universities over the last two decades. The NAS whitepaper cited two such examples – the Penn Science Teacher Institute at the University of Pennsylvania, where the author served as the founding director of this institute, and the UTeach program pioneered by the University of Texas at Austin. The PSTI programs aimed at enhancing the content preparation of in-service teachers, while UTeach, operated through the College of

Science of University of Texas, provided a new paradigm much in line with the model outside of the US for training new teachers.

The primary activities in the PSTI were the two master degree programs – the Master for Chemistry Education and Master for General Science Education. Each containing 10 courses (8 in content and 2 science education), the former aimed at in-service secondary school chemistry teachers and the latter in-service middle school science teachers. The courses were delivered over weekends of two academic years plus weekday mornings in 3 summers. The participating teacher-students were supported primarily by a grant from the National Science Foundation Math and Science Partnership program complemented by industry (Rohm and Hass was the biggest donor) and Penn scholarships. The MCE program ran 10 cohorts with about 200 in-service teachers graduated. The in-service teachers were mostly from the tri-state area near Philadelphia though many were from outside of driving distances, including one from as far as Oregon – an illustration of the needs for such programs across the country. Studies conducted in association with the PSTI programs indeed have shown that students in classrooms of the content-prepared teachers, the PSTI graduates, do learn better.

The UTeach program took in science and math undergraduate majors and provided them with early classroom teaching experiences. Students who intended to continue pursuing a teaching career would then take pedagogy as a minor along with teaching internships. Upon graduation, the student received a bachelor degree in a science or math major and a teacher certification from the State. This program, with the financial support of Exxon Mobile through the National Math and Science Initiatives, has been reproduced in similar forms in more than a dozen universities nationwide. At Temple University, the TUteach program started in 2008 as a collaborative effort between the College of Science and Technology and College of Education. By now there are about 200 students majoring in Biology, Chemistry, Computer Science, Earth Science, Math, and Physics enrolled in the program. So far about 20 have graduated with a science or math major and teaching certificate. Approximately 1/3 of the graduates are teaching in secondary schools, 1/3 working in industry and 1/3 enrolled in graduate and medical schools.

Programs like the Master for Chemistry Education and UTeach should be replicated amass in order to exert fundamental changes in the preparation of STEM human power in the US.

Innovations in higher education should also be pursued in order to prepare science students to be more industry and society relevant. The chemical, materials, pharmaceutical, and biotechnology industries in the US are experiencing significant forces of change in the new global marketplace. Scientific discoveries and new technologies continue to emerge, especially at the boundaries between traditional disciplines. Industries all over the world are required to innovate and adapt at an unprecedented pace in order to maintain their competitive edge in their field or to develop into successful enterprises. The challenges that these chemistry-related industries face demands a highly-knowledgeable workforce, especially at the leadership level, and create opportunities for higher education institutions to re-examine the preparation of science professionals, particularly through graduate programs.

There is a growing realization within chemistry-related industries of a "disconnect" between basic research activities and the successful delivery of products to the marketplace. This disconnect is revealed in several ways. Often, a technology or product is developed at the top of the research and development "stream" within a large corporation, but the ensuing effort is not effectively performed to bring the technology or product to fruition. In smaller companies, many important ideas for new technologies or products are not successfully translated to the marketplace. These failures reflect a lack of effective coupling between product development, and the execution of sound business and marketing plans to bring the product to the marketplace. Consequently, a widely held perception is that across all industries, for approximately every seven new product ideas, only one succeeds. Finally, from the management perspective, an additional disconnect is created when a business leader at a technology company does not have the scientific or technological background to effectively and continuously guide advances to sustain the growth or dominance of the company.

The gap between the development of a product and its arrival at the marketplace often reflects the educational experience of scientists who enter the corporate world, and their subsequent career history. A great majority of students graduating with doctoral degrees in chemistry go directly into industry, with most of them beginning their careers in the research and development departments of large corporations, or develop products in smaller, start-up companies. Many of these research scientists subsequently pursue a business degree, either because the degree is required for advancement through the management ranks of the corporation, or because business knowledge is required for marketing a new product or technology. The uncoupling of research training and business education can lead to flawed decisions in the product development stream by the business leaders responsible for the product, and also missed opportunities for appropriate marketing of the product or technology.

Temple University's College of Science and Technology, in which the Chemistry Department resides, and its Fox School of Business and Management have collaborated in an innovative dual-degree program to address the need for a new generation of scientists with business knowledge and better communication/management skills often derived from business/management courses. The PhD-MBA program is designed to provide a new breed of business-savvy scientists who will catalyze profound change in the manner and efficiency in which new products and technologies are brought from the laboratory bench to the marketplace, and a new breed of technology-savvy business managers who will lead established technology-focused corporations with continued competitiveness and prosperity.

Temple's College of Science and Technology and Fox School of Business and Management have already collaborated to establish an Entrepreneurship Certificate Program for graduate students in the science fields. This program is designed to educate a new breed of professional scientists that are particularly well-versed in business entrepreneurship. The program includes the completion of three business-related courses, one of which is Innovation and Management of Technology, offered by the College of Science and Technology. The PhD-MBA program in chemistry and business takes advantage of the successful partnership of the two colleges in the establishment of the Entrepreneurship Certificate Program.

The dual-degree program involves concurrent studies towards a doctoral degree in science such as chemistry and master's-level coursework in business administration. The chemistry and business course schedules are mixed in with the research activities, such that the program can be completed within 6 years, which is comparable to the time to receipt of a doctoral degree in Chemistry. Overall, 13 MBA courses are spread out over 5 years for the students to take.

Despite student interests, financial supports as well as the attitude of PhD thesis advisors toward the business courses remain major obstacles for the program to become widely taken by students. Currently it is customary for the chemistry doctoral student to be supported by a combination of teaching assistantship and graduate research assistantship derived from the PhD thesis advisor's funding. Students in this program, however, will not teach beyond their first year of study to ensure that ample time-effort in years 2-6 can be devoted to chemistry research and business courses. Whether the student is able to take the business courses during the time the student is supported to do research becomes a challenging issue unless additional resources can be identified to supplement the research funding.

There are other potential dual degree programs that can in principle be designed to fulfill specific needs of skills. For example, the ability to manage information technology is becoming a major advantage for many science professionals in completing their assigned tasks. Combining a Master of Information Technology with a science PhD program provide the opportunity for educating unique science professionals that may gain advantage in competing for employment.

Globalization not only brings challenges, but also opportunities for higher education. Many American students now pursue study abroad and many foreign students come to the US to study. To US universities the presence of international students provides diversity to the student population, enriches campus culture, and, at a time when many US universities are facing serious fiscal challenges, brings additional tuition revenue. Consequently, US universities have the obligations to tailor their curricular to suit the needs of international students. One such program at Temple University is the 3+2 Dual Bachelor-Master Degree program in collaboration with a score of Asian universities.

This program is conceived based on several considerations: Many top students in Asian universities by the time they complete the junior year, they have taken sufficient courses in meeting the admission standard of US graduate schools. Also, many Asian students enter university through a nationwide college entrance examination system in which each student is assigned to a particular university and, most importantly, major for study. It is not infrequent that a student pursues a major of study that in later days turns out not in line with the student's interest. Additionally, a student may feel the need to take on a different major for career enhancement purposes. Consequently in this DMBD program, a student from an Asian university after three years of study may enroll in a Temple master degree. During the two years at Temple, the student takes all the master courses some of which can be used to fulfill the requirements of the bachelor degree at her/his home institution. Students are allowed to change to a discipline different from her/his original major so long as all requirements can be fulfilled within two years. Many students indeed switch to a different discipline. Often engineering majors would take a finance master, math economics, or chemistry biomedicine, for example. So far more than one hundred students from China, Korea, and Taiwan have partaken in this program at Temple University.

Dual degree programs can also be used to benefit research. Research can be competitive as well as collaborative across national boundaries. In an era of rapid, convenient and low cost communication, many researchers across the world in different countries are trying to solve similar problems. Often, these laboratories have different expertise as well as instrumental capabilities. It is also the goal of national organizations such as the National Science Foundation in working with its counterparts in other countries to encourage individual investigators to collaborate to work toward finding solutions of common problems. At the university level, dual degree, in this case, dual PhD degree, programs may be an effective means.

Recently Temple University established dual PhD degree programs with the University of Belgrade in the area of computer science and Yonsei University of Korea in physical sciences. A student from one of the partner universities during her/his PhD study period come to Temple (or vice versa) to do research for 1 or 2 years. Upon completion of the PhD requirements, the dissertation is examined at both institutions. Upon passing the examination in both universities, the student receives degrees from both Temple University and the partner university. The degrees are issued individually and separately from each university so no new degree program is established. The critical factor for the success of this program is not fiscal, but rather, two principal investigators, the thesis advisors of the student, involved from the two partnering universities have to find sufficient common interests in research for the collaboration to work. The benefits on the other hand are sharing different expertise and instrumentation for solving the same set of problems.

In summary, the continuing demand of technology renovation and globalization as the new driving force for economic and social changes have presented new challenges as well as opportunities for US institutions of higher learning. Universities today must adapt to help industry and society in general to take on the opportunities and meet the challenges. New paradigms of education must aim at improving pre-college math and science education, preparing science professionals with unprecedented set of skills, and forging international collaborations.

Chapter 20

Chemistry Turns Eastwards:
A Decade in Singapore's R&D

G. Julius Vancso*

**University of Twente, MESA+ Institute for Nanotechnology,
7500 AE Enschede, The Netherlands
*E-mail: g.j.vancso@utwente.nl**

Since gaining its independence in 1965, Singapore has become one of the most competitive economies of the world and has seen enormous progress also in R&D. The author of this contribution has been part of Singapore's R&D life for more than a decade. In this contribution he tackles issues from the viewpoint of a visiting scientist, particularly with respect to the benefits and challenges of extending science using visitor's schemes and running satellite research programs in other countries. International collaborations are desirable but can be hampered by national interests, IP concerns, administrative hurdles, and other limitations. Cultural differences also exist, and concerns at home regarding loyalty, productivity, and occasional lack of appreciation must also be addressed. Nonetheless, intellectual cross-fertilization, enhanced productivity, and the benefits of nurturing and educating a new generation of "global" scientists are among the positive factors. These efforts may also lead to mutually beneficial innovations and commercial programs, using networks and local knowledge to succeed. As an example of scientific synergy, a case of global importance will be presented involving the science and international implications of marine fouling and the implementation of biomimetic strategies to avoid it.

Introduction

Chinese achievements in science and technology were in advance of Western societies for many centuries. This historical fact remained for a long time unrecognized by the West and not until the Renaissance has the West assumed leadership (*1*). Caught between historical China and prosperous Western powers are young South Asian nations, like Singapore. Since gaining independence in 1965 Singapore has moved from third world status to first, and has recently taken the second place behind Switzerland on the global list of most competitive economies of the world (*2*). Strong focus on higher education, scientific-technological research, use of scientific-technological innovation in the economy, and retaining as well as attracting talented scientists in a holistic approach have been among the major pillars ensuring this success.

The author of this contribution has witnessed the enormous development of Singapore's R&D life for more than a decade, first as science advisor, later as visiting investigator. In this chapter he will tackle issues from the viewpoint of a visiting scientist, particularly with respect to the benefits and challenges of extending science using visitor's schemes and running satellite research programs in other countries. Science is international, and scientific collaborations should transcend national borders. Yet due to political boundaries and national interests, institutionalized collaborations often encompass complex IP protection schemes, administrative hurdles, and other limitations. Cultural differences also exist, and concerns at "home" regarding loyalty and productivity, as well as occasional lack of understanding, appreciation, and sometimes even jealousy, must also be named among the challenges (*3*). On balance, however, intellectual cross-fertilization and scientific inspiration, enhanced productivity, as well as the benefits of nurturing and educating a new generation of scientists with appreciation for cultural differences must be considered. When mature, these young intellectuals, who have been exposed to different countries and cultures, will likely lead mutually beneficial science-technological and industrial programs in their home countries, and will use their broad network with roots in collaborative programs, as well as local knowledge of their remote partners, to succeed in mutually beneficial enterprises. Networks emerge and grow, which enhance the chances to successfully tackle global challenges in chemistry and in other areas of science, as well as in technology and industry. For the visiting scientist, personal benefits include a chance to enhance productivity, multiply options, place his/her students in other R&D cultures, and (last but not least) become exposed to new inspirations.

In the following narrative a personal, and by no means complete, account is given from the author's perspective about a visiting scientist's life during his research stays in Singapore while keeping his home base in the Netherlands. The author during his career crossed the Atlantic twice and lived in six countries for longer than a year. He has seen many advantages and disadvantages of scientific globalization and can speak about globalization based on his own "hands-on" experience. He is a Hungarian born scientist who was educated in part in Budapest, then continued his education and began to unfold his research career in Switzerland, immigrated to Canada and became a tenured professor at the University of Toronto, and returned again to the "old world", as holder of a Chaired Professorship in the Netherlands. As Visiting Principal Scientist, he has been participating in Singapore's science and technology life since 2001. He has largely benefitted from research collaborations and learned to appreciate the benefits of international networks. As an example for scientific synergy, a case will be presented from the portfolio of the author, involving the science and global implications of marine fouling, and the implementation of biomimetic strategies to avoid it. Finally, a few recommendations for policy makers will be presented from the author's perspective that may serve to further enhance the highly productive global collaboration schemes for the benefit of all.

The Life of a Visiting Investigator in Singapore

Numerous reports exist discussing the swift growth and the progress of R&D in Singapore, comparing key indicators, and analyzing the findings. Singapore's globalization strategy and the achievements for the first 25 years of the history of this young nation have been nicely summarized by Lai-To Lee (4) and others. Lee concludes that *"...physical size and natural resources will become less important for economic growth when compared with human capital, information, and knowledge in the future."* Indeed, the labor-intensive era of the 1960's has grown and evolved to the technology-intensive period of the 90s, which is being transformed to the knowledge-intensive decade of the present years. Singapore has moved from a third world status to the first and places strong emphasis on globalization and liberalization of its economy. There is no doubt that a smart Singapore policy to provide optimal conditions for benefitting from globalization in science and technology has made substantial contributions to the Singapore success story.

The swift development that has taken place in Singapore would not have been possible without attracting and employing foreign talent to educate Singapore's own human capital in R&D and technology. Some of the science visitors stayed for longer periods, some left and returned to their home countries, but their impact can strongly be felt and will remain visible for the foreseeable future. The high living standard, the enviable working conditions in R&D, the vibrant intellectual and multicultural life, and the ever present techno-friendly atmosphere and broad appreciation of R&D have attracted many of the world's top scientists. Yet, for a continuing success a pool of talented, receptive, motivated and willing locals is also needed to absorb the knowledge and carry it forward. Thus this issue is receiving increased attention, occasionally making foreign scientists feel a bit uneasy. Yet, still wherever the author goes in Singapore, with whomever he speaks (from a local taxi driver, or a cook in one of the popular local food courts, to a bank clerk or a company manager), when he mentions that *"I am a visiting scientist"*, the first reaction is almost always appreciation of science, appreciation of the author's presence, and expressing support for Singapore's science and technology efforts. In contrast, the author remembers when he started his first faculty position in North America and wanted to buy a car (after admitting that he was a university professor), the first reaction and response was a "gentle" inquiry about his credit status. (Apparently he did not look creditworthy enough.) When he told the car dealer of the first dealership he entered that he was European, and worked as a university professor, the car dealer looked at him and said: *"My friend, you are now in North America. Nobody cares about your brains, neither do I; all I care about is your checkbook"*.

Policy shifts and changes are sometimes needed to enhance progress. In Singapore, with a centralized political system, once such decisions have been taken, they are relatively easy to implement. In recent years, scientific objectives have been more aligned with economic reality. Application and knowledge utilization plans are also needed now as essential parts of research proposals (5) in order to support a dynamic economy and "to ensure sustainable growth". With this regard Singapore is not an exception. Yet enough room for basic science remains, and the "Singapore experiment" in science and technology, including public support and recognition, continues to blossom.

In Singapore research funding comes from various agencies, and it is not the purpose of this article to provide a detailed account of the Singapore funding system. Yet it should be noted that the main sources of funding include programs of the National Research Foundation (NRF) which supports new initiatives to develop new growth areas (defined by the government) and from the Ministry of Trade and Industry. Academic research funds are provided by these organizations (among others) to the two major public Singapore-based universities offering programs in science and engineering, i.e., the National University of Singapore, NUS, (Times Higher Education ranking for 2012-13 is #29) and the Nanyang Technological University, NTU, (Times Higher Education ranking for 2012-13 is #86), as well as funds for the research institutes of the Agency for Science, Technology and Research (A*STAR). A measure of a country's research activity is the ratio of the Gross Domestic Expenditure on R&D vs. GDP. Singapore's objective is to reach 3.5% in 2015, and it is on its way to achieve this goal.

As a leading government R&D Agency, A*STAR was established in 2002, and since then it has developed into a research organization with global importance and presence. Its principal mission is to foster world-class scientific research and talent for a vibrant, knowledge-based Singapore. A*STAR currently oversees 14 research institutes, 7 consortia and various centers located in the Biopolis and Fusionopolis complexes (see Figure 1) and in other places, for example, on the campus of NUS or NTU. A*STAR also supports research in collaboration with universities, hospital research centers and other local and international partners (see http://www.research.a-star.edu.sg/static/about) and runs a unique and truly outstanding scholarship program for incoming and outgoing students. It must be stressed that this scholarship program has been vital for A*STAR's success in attracting, nurturing, educating, and employing young talents from within the country and originating also outside of Singapore. To become a successful applicant, excellence and scientific potential are the determining requirements.

The author began his association with Singapore as member of the Scientific Advisory Board (SAB) of A*STAR's Institute of Materials Research and Engineering (IMRE) more than 10 years ago. There are still a large number of various committees, advisory and evaluation panels, international assessment boards, etc. in Singapore similar to this SAB, as these have been used to provide feedback by panels featuring external, unbiased panel members to assist policy makers and executives with independent opinion. Appointments to serve on these committees are usually for a limited time only (just like in government leadership positions). During their tenure SAB members often have a chance to establish strong contacts with local scientists and science executives. As a result, collaborations can be established and SAB members (like the author) can occasionally continue to maintain presence as visiting scientists.

Figure 1. Left: Singapore's Biopolis, which is A*STAR's futuristic international R&D hub for biomedical sciences. (Source: Wikimedia Commons.) Right: The twin towers of Fusionopolis, which is an R&D complex designed to host many of A*STAR's computer science, media, physical sciences, and engineering programs. Private companies are located within close proximity to boost application driven research and utilize results. (Author's photograph.)

The author started his visiting research program which was initially funded by the Executive Director of IMRE. For such programs no lengthy proposals were required, no lobbying needed; a good idea and a match with the Institute's mission were often enough. A short 2-3 page write-up as research plan describing the problem statement, justification, and a short summary of resources was needed for consideration. The feedback was instantaneous, and although programs like this were subject to local administrative procedures (which could be rather cumbersome, involving minute details that often slow down progress), visitor's research programs (including the author's) were generously supported. These programs had numerous benefits for both the visitor, as well as the local scientists as sketched below.

Two researchers were employed initially from the author's home institution in the Netherlands (both were looking for suitable jobs in Europe before their Singapore engagement, but could not find anything comparable) to carry out the author's proposed project, and a very fruitful period of research started. Dozens of highly cited papers have resulted; and a large number of students from the author's Dutch university, the University of Twente, followed the initially hired two "semi permanent" researchers for shorter or longer research stays. Foreign talents still have opportunities to spend research leaves or internships in A*STAR's institutes or at one of Singapore's universities (supported, for example, by A*STAR via the earlier mentioned various scholarship programs), and young Singaporean scientists often visit overseas institutions, including the author's laboratories in the Netherlands, on scholarship support. The flexibility, the resources, and the funding level were simply not provided by any European funding sources to get the work done under similarly good conditions, which was definitely a strong motivating element to maintain presence in Singapore. Efficiency was further enhanced by an ingenious scheme of A*STAR's rules, which required the employment of a local "champion" as co-investigator (locally employed, fully accountable scientist) for projects, which are run by visiting investigators; thus the visitor had no administrative duties, and could fully concentrate during short visits on the scientific problems and real issues. Such schemes of course also ensured full control by the hosting institutes. With the connectivity offered by the internet in communications, by "skype" sessions, and on-line exchange of presentation slides, it is very easy to run such collaborative programs. Joint work naturally should involve short visits (for example, on a quarterly basis) to discuss results of the previous quarter, make plans for the upcoming period, submit the papers, network, and discuss research strategy for the future. In a recent interview (3) the author called this a "very surgical operation", and the conditions offered by A*STAR strongly support such approaches.

Of the author's two initial programs, one focused on Quantum Dots and their applications as, for example, on bio labels, which also complemented and strengthened local Singapore efforts that had been underway. The first two scientists "imported" from Europe to carry out work in this visiting program have grown to become well-respected team-leading scientists in Singapore; and both chose to stay. The other program initiated by the author in his hosting institute will be discussed in more detail in the next section as an example for a challenge with global implications.

Tackling the Global Marine Fouling Challenge

Global problems require global efforts and implementation of global solutions. Marine fouling is a global problem, which can serve as a showcase to demonstrate the necessity of a scientific-technological approach across national boundaries to find adequate strategies to tackle it. When synthetic materials are submerged in sea water, dissolved matter and marine organisms attach to their surfaces, and usually colonize their interface in the process known as marine fouling. This usually leads to diminishing material performance with detrimental consequences. Various strategies exist to prevent biofouling including the use of (more or less toxic) biocides and biocide-free foul release technologies, which utilize surfaces from which the attached marine organisms can be easily removed. The author has been intrigued by this problem (and its global implications) and proposed to set up a program aiming at designing and engineering biomimetic antifouling surfaces. Such surfaces in their topology, chemical composition and surface mechanical properties mimic naturally occurring marine species that do not foul (sharks, certain shells, crabs, sponges, and some other species). Initially, in a small program the author's team first looked at the first steps of biofouling from a true molecular perspective. Before marine organisms, such as barnacles, attach to a surface and begin to grow and colonize it, their larvae often "explore" the substrate, i.e., by touching it, they deposit small amounts of proteinaceous extracts, and probe the touched spot's suitability for permanent attachment; e.g., see ref. (6). An improved understanding of the molecular principles of this process was the first target of the author's visiting program.

Biofouling is a fascinating but very complex field in science, and if progress is to be made in preventing biofouling, several disciplines must work together. The chain starts with marine biologists, and a world-class team of marine scientists at the Tropical Marine Science Institute of the National University of Singapore was particularly attractive and appealing for collaborations. This team has been doing pioneering marine biology research and collaborates with the world's top scientists in lab assays and field tests to assess coatings for antifouling performance. The tropical marine surroundings in Singapore that feature an enormous biodiversity offer a perfect field test environment. Chemists are also needed for this research, who design molecules to be included in antifouling or fouling release coatings. Environmentally benign biocide technologies have already been developed by A*STAR's Institute of Chemical and Engineering Sciences (ICES) and a team of chemists was available there. The author's hosting institute (at that time), A*STAR's IMRE, provided access to surface physical tests, polymer materials knowledge, and expertise in material surface micro- and nanoscale patterning. Scientists assigned from IMRE to this program joined forces with researchers based in ICES and other research institutes of A*STAR and the National University of Singapore to form the IMAS team (Figure 2). Finally, platform erosion testing considering lifetime assessment of antifouling coatings, and scale-up issues have been tackled by the Singapore Institute for Manufacturing Technology SIMTECH of A*STAR. Thus A*STAR provided funding for this truly multidisciplinary program covering work packages in four research institutes, and acted as a cohesive agency to help integrate the antifouling efforts with other existing initiatives. The IMAS program (Innovative Marine Antifouling Solutions for High Value Applications) was born. Clearly, this is a multidisciplinary program with obvious global implications.

The problem of biofouling is highly relevant for many technology related areas of global importance. For example, large ships move and export/import water fouling organisms in their ballast regardless of their biological "home", which must be prevented. Biosecurity, which is related to this challenge, is becoming a hot issue. Release of toxic biocides must also be prevented, for which international treaties need to be negotiated, and local legislative actions must be coordinated. Solution providers for antifouling (e.g. coatings companies) must cooperate with users, such as marine authorities, navies, and underwater signaling providers, offshore structure owners in the petroleum industry, aquaculture operators, water treatment facility organizations, and so on. Last but not least, "spin-off" results of antifouling research include, for example, a better understanding and design criteria for bioadhesives, which can be applied in medicine. Thus biofouling is a truly global and fascinating problem which requires global collaborations and global solutions. Singapore provides an ideal environment for biofouling research and has substantial vested interests in using the results of biofouling research. Within one year following the start of this A*STAR funded program, 20 researchers have become engaged for IMAS, some of them were recruited from overseas countries, such as Europe and the Americas. We engaged scientists from the cadre of Singaporean researchers and made them experts, as well. Missing equipment and scientific infrastructure were acquired using funds provided by A*STAR. Upon completion of the ongoing incubation phase of IMAS, it is anticipated that a continuing activity will be transformed to a technology transfer phase to serve industrial interests, as well. Thus IMAS has indeed become a showcase for a research program of global technological implications, carried out by an international team rooted and funded in Singapore.

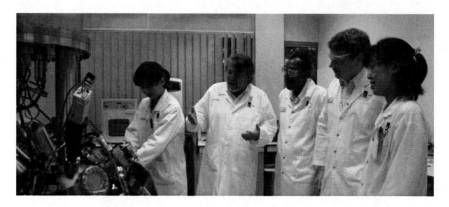

Figure 2. Researchers with international backgrounds at ICES of A*STAR investigate and discuss surface properties of a platform to be used as substrate for a model antifouling coating. The author is shown as second from the left.

Recommendations

There is a large body of literature discussing implications of science globalization. Top journals such as Science (*3*) or Nature (*7*) often devote leading scientific or editorial articles to this topic. Fairly recently, in an issue of Nature (*7*), which focused on globalization, editors and science journalists analyzed recent fundamental shifts (science "on the move") and discussed how to build science capacity crossing national borders. The authors also analyzed recent trends in "the rise of research networks" across the globe. Obviously, global challenges need global solutions. It is reassuring to acknowledge that science policy leaders recognize many of the challenges and consider solutions. However, equally important is to listen to those scholars, who are active (or managing research) at the lab floor level, and are personally involved in the globalization process, either as guests or hosts, or simply as scientists on the move. In ref. (*7*); p. 331, Mr. Lim Chuan Poh, Chairman of A*STAR proposed ways to boost research in the next decade suggesting that Singapore's way is to build global networks. It is hoped that this contribution to the ACS Presidential Symposium has provided ample evidence that indeed going global in science and building of such global networks is a preferred way to go.

Before concluding, the author would like to recommend the following points for consideration:

- Data on globalization in science is patchy and often anecdotal; in the literature mostly ad-hoc studies are available, and it is difficult to find reliable information. National societies may wish to conduct coordinated surveys, monitor/publish globalization trends for increased awareness, and publish these in high-circulation magazines (e.g., C&EN).
- For success, continuing education is needed to tackle cultural and language barriers. Understanding of respective needs and knowledge of local culture are indispensable for success. Mutual respect (by the hosts as well as the visitors) is important, and one should not consider imposing his/her own system on the others once a collaboration has been started. Human diversity (like biodiversity) enriches: there is no such thing as one size fits all.
- Global science needs global peer reviewers. Scientific ethics should never be compromised, and should also become global. Science leaders should promote high, unified scientific standards in publishing and reviewing across the globe.
- Mutually advantageous funding schemes are needed for collaborating parties. As an example, A*STAR's Graduate Academy scholarships for Singaporeans and foreigners can be considered. Sandwich PhD programs are highly recommended, and simplified funding schemes should be introduced to promote these options.
- Bureaucratic hurdles must be reduced to the absolute necessary, bare minimum (on all sides).
- National policy makers may want to select a few focal areas in chemistry that have global impact. They should place these topics on political agendas, promote, and fund the corresponding research programs nationally and globally.
- Running satellite labs is a good option, but it is very challenging and demanding on the participating individual. For such "global" scientists, the home base may become compromised, while appreciation from administration and colleagues is often limited, for example due to lack of understanding. Enhanced awareness of the benefits of these programs for everyone, and corresponding appreciation would be beneficial.
- How can ACS help? Enhance publicity of global collaboration. Publish topical C&EN several times a year devoted to global chemistry. Consider creating prestigious awards to recognize globalization efforts.

Acknowledgments

The author is truly grateful to Dr. Marinda Li Wu, ACS President, for the invitation to attend and lecture at her Presidential Symposium at the 245th ACS National Meeting in April 2013 in New Orleans. Dr. Wu has made globalization as one of her presidential themes and the author wishes her and ACS great success to reach their objectives. Thanks are also due to Dr. H.N. Cheng of the USDA Southern Regional Research Center for his great organizational efforts with respect to this symposium, for his help with writing this chapter, and for his friendship during the last decades. During the last 10+ years in Singapore the author has immensely enjoyed the challenge of working among the very different South-East Asian boundary conditions, strengthened his interpersonal skills by doing research and becoming successful in a different culture; enjoyed science and scientific productivity, and also enjoyed Singapore's lifestyle. Asia and the Asian culture provide endless inspiration, enjoyment and pleasure. A large number of people (too many to list) have helped the author, and he wishes to express his gratitude for the support, collegiality, and friendship to his colleagues in Singapore, in particular at A*STAR, who have made his "global journey" possible.

References

1. Needham, J. *The Grand Titration: Science and Society in East and West*; University of Toronto Press: Toronto, Buffalo, 1979.
2. *World Economic Forum, The Global Competitiveness Report 2012-2013.* http://www3.weforum.org/docs/WEF_GlobalCompetitivenessReport_2012-13.pdf.
3. Service, R. S. News Focus: Satellite Labs Extend Science. *Science* **2012**, *337*, 600–603.
4. Lee, L.-T. Singapore's Globalization Strategy. *East Asia, An International Quarterly* **2000**, Summer, *18* (2), 36–49.
5. Editorial. Singapore's salad days are over. *Nature* **2010** December 9, *468*, 731.
6. Phang, I. Y.; Aldred, N.; Ling, X. Y.; Huskens, J.; Clare, A. S.; Vancso, G. J. Atomic force microscopy of the morphology and mechanical behavior of barnacle cyprid footprint proteins at the nanoscale. *J. R. Soc., Interface* **2010**, *7* (43), 285–296.
7. Special section, articles. The New Map of Science. *Nature* **2012** October 18, *490*, 325–338.

Chapter 21

Young Scientists without Borders

Jens Breffke*

Department of Chemistry, The Pennsylvania State University,
University Park, Pennsylvania 16802, United States
***E-mail: jens@psu.edu**

In today's global chemistry enterprise it is important for young scientists to have not just the best scientific skills but also the soft skills that gives them the advantage over their competing job seekers and coworkers. Mastering team and project management today also requires the ability to deal with ethnic diversity across cultural values and language barriers. Therefore the need for opportunities for young scientists to experience diversity becomes a key element of success. There are exchange scholar programs run by governmental institutions. In addition, scientific non-profit organizations with their younger member committees can often provide additional opportunities very effectively. The author currently presides the International Activities Work Group of the ACS Younger Chemists Committee. A native of Germany, he was chair of the German JungChemikerForum of the German Chemical Society. He promoted the idea and founded in 2007 the European Younger Chemists Network (EYCN) of EuCheMS. EYCN today is an association of 25 European younger chemist divisions within their societies, representing more than 50,000 younger chemists.

The globalization of the world's leading industries is well underway, and by 2025 most countries will be working intertwined to perform research, manufacture products, and distribute those products to consumers around the world. The chemical industry is one of these market segments where globalization has produced great rewards with cheaper products and innovative technologies incorporated into sophisticated, high-tech products. This trend will continue which demands additional and inevitable skills from the people working in this field. The working environment becomes more and more diverse with respect to culture, language, and religion. With international colleagues entering your home workforce brings challenges, but the chances that you already are or will be the foreigner yourself in someone else's country are getting higher.

During my tenure as chair of the German JungChemikerForum (JCF) of the Gesellschaft Deutscher Chemiker (GDCh) in 2006/07, I was a student at the Humboldt-Universität zu Berlin in Germany working in a research group side by side with people from Russia, Spain, India, Iran, and China. I found that each nationality has its own humor, habits, principles and predominant religions. These are each important to recognize and embrace. For instance, if you are not familiar with other religion's practices, it may be surprising to find your Muslim co-worker praying inside the lab behind the bench on the floor in the middle of day on a small carpet. I was always fascinated by these kinds of things that you just had never seen or experienced before.

At this time in my studies, I had the great pleasure to invite my counterparts from other European chemical societies to come to Berlin to work out the by-laws for the foundation of the European Young Chemists' Network within the European Association of Chemical and Molecular Sciences (EuCheMS). Together with my Hungarian fellow Csaba Janáky, then student at the University of Szeged, we worked at this meeting to bring all people to an agreement that satisfied all. Thirteen societies were represented by about 20 delegates that had mostly never seen or even really talked to each other. Looking back today on why this meeting was a success despite the tremendous challenges in moderating such a group, then because we were able to think beyond what we were personally used to and believed to be right and good and to respect what other people had to say. This perspective goes along with the patience to listen to someone who doesn't speak the meeting language very well. Today EYCN works for the benefits and interest of more than 300,000 younger chemists in more than 20 societies in Europe. This is possible only because of the attitudes of respect and openness on the part of the leaders involved.

Since 2001, the German Chemical Society (GDCh) has had an ongoing exchange program with the Northeastern Section of ACS (NESACS) where younger chemists travel to an international conference in the respective other country. As a selected delegate I was granted the opportunity to attend the ACS National Meeting in Boston in August of 2007. Having worked with chemical societies in Europe and their conferences it was a completely new experience seeing the scale of operation of the American Chemical Society. This exchange program includes participation at a national or regional meeting and is followed by a professional as well as a social program. Visiting large pharmaceutical companies and labs at Harvard and MIT in combination with a trip to Cape Cod left impressions for a lifetime.

After returning to Germany I was never able to completely let go of the idea of potentially pursuing a PhD career in the USA. But the truth is I wasn't sure if I could really handle it. Am I good enough? Would anyone in the States be willing to take me as a graduate student? Could I handle leaving my home country for an undetermined amount of time? Maybe I would never come back!? Even though I enjoy and embrace diversity I was not free of doubt. My mind said it would be great but my heart had doubts. I talked with my research mentor, who was from Spain, and he offered to connect me to his old PI and in order for me to possibly go abroad for three months to Spain. What sounded like a great idea turned out being more involved than I initially anticipated. Fortunately, I was able to get a Leonardo-da-Vinci II Fellowship by the German Academic Exchange Service (DAAD) that made it affordable – and it was certainly a big improvement to my CV. The most distinct experience I had by going to the University of Santiago de Compostela in the northwest of Spain was something I didn't quite expect: People there barely spoke any English – and I didn't speak any Spanish. I thought that in a college town people would easily speak English but I was mistaken. During this one semester in Spain I acquired a basic knowledge of the language without attending any actual classes. The take-home message for me was this: Even though I didn't speak the language I had a great time and was able to do some interesting science even though the lab I worked in was not equipped as well as I was used from Berlin. Back in Germany I knew I was able to handle being long distances from friends and family, the change in language, and being able to adapt to a new environment.

That being said I decided to send out applications to professors in the United States for whom I wanted to do research to get my PhD. That would have been the way to go if you would want to get into a research group in Germany. However, something that I was not aware of is that in the US you have enrollment into graduate school generally only for the fall semester and that the GRE test and, for foreigners, an additional TOEFL test are required. My anecdote to the GRE General: The verbal part is meant to challenge native speaker so it seems unrealistic that a non-native speaker could ace it, but that's why you have to take the additional TOEFL test to somehow make up for that. I do agree though that I should be able to master the math part as someone who is in the STEM sciences. However, I performed only moderately on that part and here is why: I am scientist. We work based on the SI unit system. The GRE math part is asking you: "You have a bathroom that is 5x7 ft. Your tiles are 11x11 in. How many tiles do you need to cover your bathroom floor?" I was honestly not prepared for that kind of question. Not that I could not have answered the actual geometry question but the conversion from foot to inch was unknown to me – at the time. Later in an interview with graduate schools I was asked why they should accept me with such a poor performance in math according to the GRE math test. Fortunately, after I told this anecdote to the committee they granted my application while laughing loudly. But that story could have gone poorly for me, too.

Despite all the administrative challenges I got lucky by writing a professor at Penn State who I knew from a visit at the Humboldt University a year earlier and who did similar research to what I worked on. Thirty-six hours after I sent him my application by email he replied positively to my request. Due to the different timeline of German semesters I was not able to get into the graduate program for fall of 2008; however, my new PI made it possible for me to be admitted for the spring semester of 2009 which is when I started graduate school at the Pennsylvania State University. Two major problems occurred with respect to the transferability of my German Diplom degree to the US. The German Diplom Degree is usually a 6-year program and generally considered Masters-equivalent. However, my transcripts were available in German only and my program didn't use the credit point system. As result no US university would have accepted my degree as a Masters equivalent degree. This problem is one of the major reasons why going abroad was not very appealing to Germans at the time. A lot has happened since then in the German educational system and today programs do have English syllabi and provide a credit point scores.

Starting over again and having to take classes for one more year is certainly not appealing, but one has to keep the bigger picture in mind. After finishing my involvement in the JCF and EYCN I focused on getting settled in the United States. Not long after I arrived at Penn State I received a message from a former chair of the ACS Younger Chemists Committee, Melissa Hellman. We met in 2008 at the EYCN Delegate Assembly in Madrid, Spain. We had invited an YCC representative to join our meeting and offered affiliation to establish long term communication channels. In this message she encouraged me to submit my resume to the ACS committee for appointment. This was an opportunity I couldn't let pass. My appointment came right away for January of 2010. Mick Hurrey, then the chair of YCC, asked me to take initiative on some international programming for the upcoming International Year of Chemistry (IYC) 2011. Two events were successfully organized. First, we coordinated a symposium at the National Meeting in Anaheim in April 2011 in which speakers from North America and Europe presented remotely to talk about different graduate school opportunities in different countries. In this symposium it was important to me that the presenters explicitly addressed degree transferability and other administrative obstacles of which one should be aware. Second – because I am a child of the program by NESACS and GDCh – I proposed that YCC launch an exchange program based on the same template. This was accomplished, and we had six European younger chemists and a representative from EYCN come to the ACS National Meeting in Denver. This pilot project required substantial financial resources and we were able to acquire all funding necessary to make it happen. With the support of the ACS Denver local section, European students from Finland, Russia, Spain, Germany and Austria had similar positive experiences like what I experienced just a few years earlier. Even past IYC 2011 YCC was able to continue the true spirit of the exchange program and US students went in 2012 to the European Chemistry Congress in Prague, Czech Republic.

Summarizing, certainly not all events went smoothly and did not just happen. But this is what it is about: You grow with the challenges – not with the easy parts in life. I encourage everyone to look for opportunities and make them happen. Rarely these opportunities will come to you by the themselves or will reveal as such easily. In order to succeed you may have to accept set backs in the beginning.

In ancient Rome they labeled maps with "hic sunt leones" (here are lions) in areas that were unknown. The bottom line is globalization is already here and there is no way back. Globalization can be embraced and should be embraced – not feared. One needs to be open minded to the things we don't know and are not used to even though they may be strange to us.

Chapter 22

Vision 2025: Helping ACS Members Thrive in the Global Chemistry Enterprise

Marinda Li Wu

2013 President, American Chemical Society,
1155 Sixteenth Street N.W., Washington, DC 20036, United States
***E-mail: marindawu@gmail.com**

This is the final report of ACS Presidential Task Force on "Vision 2025: Helping ACS Members Thrive in the Global Chemistry Enterprise." The initial Presidential Task Force report with its seven recommendations was first shared with the ACS Board of Directors in December 2012 . Pertinent recommendations were then shared with 27 stakeholder committees and divisions of ACS to gather more feedback for refinement of the recommendations. The final Presidential Task Force report was issued in June 2013 and disseminated within the ACS. The report has been slightly adapted for publication in this book -- the *ACS Symposium Series* 1157, published in 2014.

Introduction

At the spring 2012 ACS National Meeting in San Diego, I held the inaugural meeting of my Presidential Task Force "Vision 2025: Helping Members Thrive in the Global Chemistry Enterprise." It has been quite a journey, and I want to express my heartfelt thanks to the entire Task Force for their commitment and hard work to make this report possible. Their names and biographies are given at the end of the report.

I especially would like to recognize the dedication of my two co-chairs, H.N. Cheng and Sadiq Shah, and Bob Rich, our ACS staff liaison. I want to also acknowledge the participation and help of many ACS committees and divisions we have consulted along the way. Last but not least, I wish to acknowledge the wise comments from many friends and colleagues who have always expressed their desire to show support. Wthout all their contributions, hard work and commitment, this report would not have been possible.

We hope that this report will stimulate productive discussions and engage both ACS leaders and members through 2013 and beyond.

As we all know, our members have been especially hard hit by the continued mass layoffs in the chemical industry and particularly in the pharmaceutical industry. This report summarizes the Task Force recommendations to address some of these challenges we face in the current global chemistry enterprise.

In my view, it is critical to continue working together to address these challenges and turn them into opportunities. My Presidential Theme for 2013, "Partners for Progress and Prosperity," highlights actions to address the challenges and take advantage of the opportunities.

Happy reading. If you have comments, please email me at m.wu@acs.org.

I. Executive Summary

Globalization of the chemistry enterprise has posed a myriad of tough challenges for ACS members. To address these challenges, Dr. Marinda Li Wu, ACS 2012 President-Elect, appointed a Presidential Task Force "Vision 2025: Helping ACS Members Thrive in the Global Chemistry Enterprise." Task Force goals include identifying challenges and opportunities, and helping members to find jobs and thrive. The Task Force has studied the issues and is providing the following recommendations for ACS leaders, volunteers, and staff:

1. Better educate ACS members about the critical elements necessary for success in a broad spectrum of career paths.
2. Strengthen ACS efforts to support entrepreneurship.
3. Engage and equip members with enhanced advocacy tools and training so that they can proactively contact their legislators to improve the business climate and aid jobs creation.
4. Explore with U.S. and global stakeholders the supply and demand of chemists/jobs to bring them to a better equilibrium.
5. Collaborate with others, including chemical societies around the world regarding public communication, education, advocacy, chemical employment, and other topics of mutual interest.
6. Provide information, resources, advice, and assistance to ACS members interested in global job opportunities.
7. Expand ACS support for chemists and chemistry communities worldwide.

In addition, the Task Force has initiated actions on the following:

1. International Employment Initiative (IEI) launched at the 2013 spring ACS National Meeting in New Orleans
2. Global Opportunities Symposium and Global Collaboration Roundtable Discussions with chemistry enterprise leaders worldwide which took place at the April ACS National Meeting in New Orleans

3. Career Advancement Opportunities Symposium, Innovation and Entrepreneurship Symposium, Impact of Diversity and Inclusion Symposium, and Advocacy Training Workshop to take place at the September ACS National Meeting in Indianapolis

II. Detailed Recommendations

The Task Force has studied the issues and is providing recommendations for ACS leaders, volunteers, and staff. Most of the recommendations are meant to build upon and supplement existing successful programs (*1*). The recommendations are to

1. Better educate ACS members about the critical elements necessary for success in a **broad spectrum of career paths** (*2*).

    A. Raise member awareness of the availability of **massive open online courses** (known as MOOCs), including simplified short lectures on technical and business skills, where large numbers of people can attend the webinars (*3*). There are widely different views on the usefulness of MOOCs which have recently received much attention.
    B. Make available recorded **web-based** career development/ management workshops on different topics and new technology areas. Provide online "on-ramps" for different groups to access relevant content (*4*).
    C. Expand online video and other vehicles to help **empower members in their job searches**, especially focused on targeting the word-of-mouth job market (*5*).
    D. Graduate and undergraduate **curricula** should include technical skills, career management skills, interdisciplinary and business training, communications skills, and some knowledge of cultures worldwide. Such courses can also be suggested to supplement ACS Leadership Development courses.

2. Strengthen ACS efforts to **support entrepreneurship** (*6*).

    A. Find ways of **linking ACS entrepreneurs with venture capital funds** and angel investors (*7*).
    B. Make recommendations to graduate programs to include topics such as **intellectual property protection and commercialization** in their curriculum (*8*). Also, offer workshops on intellectual property at ACS National Meetings (*9*).
    C. Offer **entrepreneurship educational** programs to showcase and support business start-ups. Create programming to support a network of entrepreneurs (*10*).

D. Facilitate and explore programs to change the culture so that more chemists are encouraged to pursue the path of **entrepreneurship** (*11*).

E. Build an ACS "brand" around supporting chemical entrepreneurship.

3. Engage and equip members with **enhanced advocacy tools and training** so that they can proactively contact their legislators to improve the business climate and to promote job creation (*12*).

    A. Provide chemists with an effective **elevator advocacy pitch** on the economic importance of increasing and sustaining federal R&D investment and improving the business climate to help with jobs creation and retention (*13*).

    B. Create **webinar-based educational programs on advocacy** in collaboration with Local Sections to make an impact at the state and local levels.

    C. Identify jobs and business growth policy initiatives and prepare members at local sections to advocate to **state legislators** (*14*).

    D. **Survey** chief technology officers and R&D leaders in companies to obtain their views on **effective advocacy in the current jobs climate** (*15*).

    E. Promote education programs to increase the competitiveness of the **U.S. workforce in areas of need**.

    F. Have a discussion with leaders at the Global Collaboration Roundtable as to possible **advocacy targets**.

    G. Improve the **business climate** for job creation. The United States needs to remain competitive and leverage the global talent pool (*16*).

4. Discuss with U.S. and global stakeholders the **supply and demand of chemists/jobs** to bring them to a better equilibrium (*17*). Initiate a task force to look at options, including immigration-related issues.

5. **Collaborate with others, including chemical societies around the world** regarding public communication, education, advocacy, chemical employment, and other topics (*18*).

    A. Actively reach out to these societies through the establishment of more **international chapters** and other on-the-ground partnerships.

    B. Establish **new awards** to support the international chemistry community.

C. Use *C&EN* to publish perspectives from officers of non-U.S. societies, and ask them to include perspectives from ACS officers in their magazines in return.

D. Support and encourage jointly sponsored **ACS-IUPAC meetings** (*19*).

6. Provide information, resources, advice, and assistance to ACS members interested in **global job opportunities** (*20*).

A. Support and promote the **ACS International Center** with its launch.

B. Make global opportunities a **national meeting theme.**

C. Provide advice and assistance to members considering **overseas assignments** within multinational companies and elsewhere.

D. Encourage more chemistry majors both at the undergraduate and graduate levels to actively participate in **study abroad programs** to better prepare for global employment, exchange programs, and future international collaborative opportunities, and for global workforce readiness in general.

E. Consider **ACS certification** of some non-U.S. chemical education programs (*21*).

F. Work to improve worldwide acceptance of chemistry **degrees and course credits**.

7. Expand ACS **support for chemists and chemistry communities worldwide** (*22*).

A. Enable overseas members to pay ACS **dues in their own currency**.

B. Continue and expand our welcoming stance toward the establishment of **new international chapters**.

C. Encourage **Technical Divisions** to expand involvement of international members (e.g., joint topical meetings, pilot grants for global programs, and jointly sponsored international meetings).

D. Plan **overseas "regional" meetings** that build upon Pacifichem's success.

E. Hold a **Virtual International Chemistry** meeting.

F. Hold more ACS **webinars with global interest.**

G. Continue work to **streamline visas** for non-native scientists to speak at U.S. conferences.

H. Expand engagement of and programs for **non-U.S. nationals** who are studying in the United States.

III. Actions Underway

In addition, the Task Force has suggested action items, which are in the process of being implemented:

1. International Employment Initiative (IEI)
   Working with staff of the ACS Department of Career Management and Development, the launch of IEI took place at the ACS 2013 Spring National Meeting in New Orleans. The ACS Virtual Career Fair (VCF) was the platform for this initiative. Prior to the IEI, most of the companies using VCF were domestic; only one international company used this service. The IEI offered international employers an additional way to connect with job seekers. The VCF permits international companies to post job openings, screen resumes, and even interview candidates online. Job-seekers can apply for openings, ask questions, and request interviews (this process can continue up to 90 days after the national meeting).

   The IEI launch was publicized with several computer terminals in a special area at Sci-Mix and job-seekers were invited to apply for international jobs. At the same time, we recruited international companies to register for the VCF in New Orleans and to post job openings. Hopefully, international companies will continue to use the VCF to post future job openings.

2. Global Opportunities Symposium and Global Collaboration Roundtable Discussion
   These two events were organized together for the national meeting in New Orleans and took place on Monday-Tuesday, April 8-9, 2013. There were three symposium sessions. The first session discussed global opportunities from thought leaders in business, academia, and government. The second session provided perspectives from presidents of eleven chemical societies representing Europe, Asia, Africa, and the Americas. The third session focused on sharing experiences from global start-ups and people with successful overseas assignments.

   At the end of each session, there was a panel discussion, where all the speakers answered questions from the audience and elaborated further on specific points. On both days, there were small group discussions (i.e., a Global Collaboration Roundtable) over a working lunch, where we exchanged ideas on collaboration, explored perspectives on the global chemistry enterprise, and formulated joint action items. The presentations from this symposium are being published in an ACS Symposium book, which will allow us to share the insights more broadly.

3. Career Advancement Symposium, Innovation and Entrepreneurship
    Symposium, Impact of Diversity and Inclusion Symposium, and
    International Chapter Summit
    At the ACS 2013 Fall National Meeting in Indianapolis, a Presidential
    Career Advancement Symposium will highlight and discuss a wide
    range of possible career pathways for chemists. A Symposium on
    Innovation and Entrepreneurship will feature various speakers who
    can discuss their experiences in starting small businesses based upon
    innovative technology. At the National Meeting, there will also be a
    symposium on the Impact of Diversity and Inclusion to highlight and
    raise awareness of various underrepresented groups of the Diversity
    and Inclusion Advisory Board. In support of global collaboration, there
    will be the first summit of ACS International Chapters at the National
    Meeting to provide guidance and an opportunity for strategic planning.

4. Advocacy Training Workshop
    Also at the Indianapolis ACS National Meeting, a special Presidential
    Advocacy Training Workshop called "React with Congress: Become a
    Chemistry Advocate" will be organized by the Committee on Chemistry
    and Public Affairs and cosponsored by the Committees on Economic
    and Professional Affairs, Public Relations and Communications,
    Younger Chemists, and Senior Chemists, and the Division of Industrial
    and Engineering Chemistry. It will help ACS members to more
    effectively advocate for economic growth through science funding,
    STEM education, and an improved U.S. business climate. Learning
    how to advocate in this way can help with job creation. ACS cannot
    create jobs, but we can help improve job prospects, if our members get
    involved in advocating with legislators to reduce barriers for innovation
    and stimulate business growth.

IV. Goals and Objectives

To guide her term within the ACS presidential succession, culminating in
service as ACS president in 2013, Dr. Marinda Li Wu has set five goals:

1. Serve our members' interests
2. Promote science literacy and education
3. Drive action, transparency and inclusivity
4. Build bridges for strategic collaborations
5. Advocate for jobs and professional growth

Since Dr. Wu has been supporting the first three goals for many years, goals
4 and 5 are the primary focus of this Task Force, *Vision 2025: Helping ACS
Members Thrive in the Global Chemistry Enterprise.* Both of these goals support
the ACS strategic goal to "Empower an inclusive community of members with
networks, opportunities, resources, and skills to thrive in the global economy"
(*23*). Dr. Wu's goal 5 also supports the ACS strategic goal to "Communicate

chemistry's vital role in addressing the world's challenges to the public and policymakers."

In support of these higher-level goals, the Task Force has been challenged to pursue the following **Task Force Goals:**

1. Identify challenges and opportunities related to the global chemistry enterprise with respect to job growth, collaboration, education, and advocacy
2. Provide recommendations to help members with jobs and to thrive in the global environment

The **Task Force Objectives** include the following:

1. Enable and enhance global collaborative partnerships and exchanges
2. Encourage business creation, growth, and expansion of the global chemistry enterprise
3. Prepare chemists for a wide range of careers
4. Engage members in global initiatives (including education, research, policy, and partnerships)

V. Process

Membership

The Task Force, appointed by Dr. Marinda Li Wu, is divided into two working groups: *Globalization Opportunities* chaired by Dr. H.N. Cheng, and *Jobs and Advocacy* chaired by Dr. Sadiq Shah. Dr. Wu leads the Task Force along with the two working group co-chairs, and Dr. Robert Rich (ACS Director, Strategy Development) who serves as staff liaison.

The Task Force appreciates the contributions of many ACS leaders, committees, and divisions as this report was developed, and looks forward to continuing collaborations as the recommendations are refined and implemented.

Structure of Work

The two working groups held face-to-face meetings at both 2012 national ACS meetings and monthly virtual meetings. They assessed the current landscape; identified gaps, threats and opportunities; and ensured that any ideas recommended do not duplicate existing ACS efforts, but will strengthen these efforts and leverage synergies that exist.

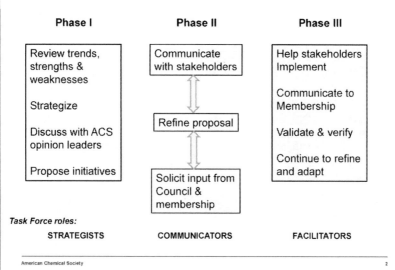

Presidential Task Force - Work Plan

Phase I	Phase II	Phase III

Review trends, strengths & weaknesses

Strategize

Discuss with ACS opinion leaders

Propose initiatives

Communicate with stakeholders

Refine proposal

Solicit input from Council & membership

Help stakeholders Implement

Communicate to Membership

Validate & verify

Continue to refine and adapt

Task Force roles:

STRATEGISTS   **COMMUNICATORS**   **FACILITATORS**

The following approach was used:

1. Brainstorm strengths, weaknesses, opportunities, and threats (SWOT)
2. Consider related external trends and leading indicators
3. Consider related existing and planned ACS activities
4. Develop possible recommendations for new and enhanced offerings
5. Consult with key committees and other stakeholders on possible recommendations
6. Present draft recommendations to ACS Board of Directors
7. Revise recommendations and share broadly to encourage implementation (to take place in 2013 and beyond)

VI. Findings

It is evident that globalization is here to stay. In the past 20 years, the number of U.S. jobs in the chemical sciences has steadily decreased. Many chemistry-based products have been commoditized, and the chemistry enterprise has been globalized. For the sake of the global chemistry enterprise and its practitioners, it is increasingly important to collaborate, to take advantage of the opportunities globalization offers, and to address the challenges it creates.

227

1. Environmental Scan

The following key trends were observed and serve as the backdrop to this report:

- The chemistry enterprise continues to be globalized. Chemical products, R&D, manufacturing, and services (as well as associated jobs and capital) increasingly move seamlessly across national boundaries.
- GDP growth is relatively low in the United States, United Kingdom, Germany, and Japan. It is high in several developing countries.
- Many chemical products have become commodities. In many industrial sectors, the United States no longer has a clear technology edge. The technologies are well known, and the major competitive advantage is price.
- According to the U.S. Bureau of Labor Statistics and National Science Foundation (NSF) data, chemical jobs have been decreasing in the United States during the past 20 years and will continue to decline in the near future.
- The average unemployment rate in the country is 7.9%, and the unemployment rate for ACS member chemists is 4.2%, which represents a decline from the previous year's all-time peak of 4.6% (24). However, this is still very high by historical standards, especially within certain specialties.
- Among new graduates, the average unemployment rate is 13.3%, compared with a record 4.6% for those in chemistry (24).
- Recently employed B.S. graduates are more likely than those with a Ph.D., or ACS members as a whole, to be employed outside their field in areas not commensurate with their education, or they are engaged in work not viewed as professionally challenging.
- Despite high unemployment for new chemistry graduates, the number of fresh M.S. and Ph.D. graduates has increased.
- The boundaries between chemistry and other sciences are becoming blurred, and there has been more research at the interfaces. Today's jobs are crossing traditional disciplinary boundaries, and we have seen increasing inter- and multi-disciplinary content in degree programs to support industry's needs.
- Students are concerned about their future in chemistry and are experiencing difficulty navigating the job market. Graduate students, in particular, are concerned about the lack of jobs in industry and available faculty positions. One faculty member put it this way:

> *"Think about it. For every 20 or so students that are graduating in chemistry, you have one, maybe two job openings. Eventually, there is going to be a glut of chemists for the few jobs that become available."*

228

- Students added that post graduate/doctorate studies have become necessary just to land entry-level positions within the private sector.

  "It used to be that places like Abbott would hire chemists right out of college (undergraduates). Now, you need, at a minimum, a Masters and it helps to have experience in more than one subfield...The job market is scary out there."

- Students recognize the need to improve skills related to job search. Students and postdocs are looking for opportunities to network among academic colleagues and those in industry who can provide jobs. Networking is among the most important benefits a professional association can offer. Many students struggle to define networking and admit to lacking this skill. There is a strong interest in ACS resources to help.
- Students and faculty note the need for better preparation for careers outside of graduate academic institutions, with which many faculty members are unfamiliar.
- There is a need to engage a diverse pool of future scientists, who can support U.S. competitiveness.
- The United States is still the envy of the world as far as graduate education is concerned.
- U.S. academic institutions of higher education are increasingly building partnerships with universities in other countries for education and research. This provides a competitive advantage and global opportunities for those U.S. graduates who have had an international exposure as part of their education.

2. *Trends and Challenges*

The Task Force identified some trends and challenges:

- There will be fewer chemical jobs in the United States in the future
- The chemistry enterprise is globalized
- Commoditization of chemical products is occurring
- There remains a persistent negative perception that the general public has towards chemicals and the chemical industry
- Boundaries between chemistry and other sciences are blurred
- Unprecedented budget deficits and demands on government finances at all levels of U.S. government (and within the European Union) constrain investment in science and engineering
- Continued population growth results in the following global challenges:

  - Depletion of earth's resources
  - Rising energy costs not likely to decrease soon
  - Increased air and water pollution

3. *Key Questions*

The key questions to address are

∎ What should ACS do?

  - What constructive role can ACS play?

∎ How does globalization affect us?

  - How can we convert challenges to opportunities?

∎ What should a chemist do?

  - Where will the jobs be in the future?

4. *ACS and the Global View*

The following diagram depicts the global presence of ACS today. At ACS, some major activities include the Chemical Abstracts Service (CAS), ACS publications, meetings, and membership. In CAS and publications, we are quite global. Our journals are sold internationally. The number of international papers and patents now surpass those from the USA. Most ACS meetings are currently located in the United States, including the national meetings, regional meetings, and local section meetings. There are some meetings (like Pacifichem) that are collaborative efforts among several national chemical societies. Some ACS technical divisions also organize international meetings outside the country. Roughly 15% of ACS membership currently resides outside the United States, and international ACS membership is growing. There are currently six international chapters with the newest international chapter in Romania approved by ACS Council at the April 2013 national ACS meeting in New Orleans. The first ACS international chapter summit is planned for the national ACS meeting in Indianapolis in September 2013.

Increased partnership and collaboration with other national chemical societies for mutual benefit can be an important key for success. These societies share common challenges and by working together, we can collectively have greater impact and make more significant progress.

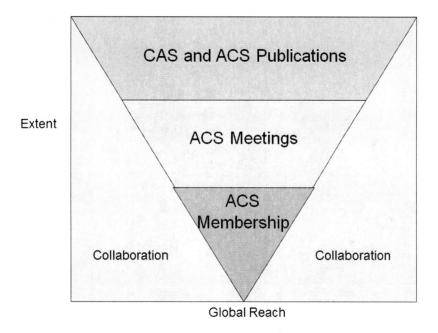

5. *Jobs and the Future*

The chemistry enterprise covers a large number of sectors, e.g., chemicals, pharmaceuticals, biomedical devices, plastics, rubber, coatings, personal care, cosmetics, process and water treatment, as well as many others. Continued collaboration is needed with and among industrial chemists to address the issues of jobs, education, and globalization. Support of small businesses and entrepreneurs should be an essential part of this effort.

At the same time, we must work to expand the universe of available career options for chemists. In the figure below, the sizes of the circles are meant to illustrate the number of jobs available in different fields of employment. As we know, the number of strictly "chemistry jobs" is limited (corresponding to the small area of the innermost circle); thus, we need to look at science-related jobs in order to increase the number of jobs available to chemists and chemical engineers (corresponding to the areas of the larger circles). The circles do not imply that the skills in different circles require less chemistry, or that the distance away from the center means diminished value. Furthermore, the scientific skills needed in technical service, management, inter and multidisciplinary research fields or other chemistry-related jobs can be just as technically demanding or rigorous and rewarding as traditional R&D jobs.

Jobs – A Simplified View

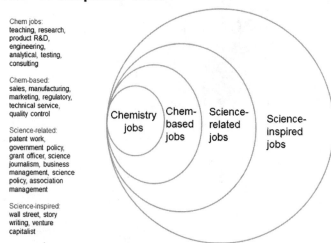

Chem jobs:
teaching, research,
product R&D,
engineering,
analytical, testing,
consulting

Chem-based:
sales, manufacturing,
marketing, regulatory,
technical service,
quality control

Science-related:
patent work,
government policy,
grant officer, science
journalism, business
management, science
policy, association
management

Science-inspired:
wall street, story
writing, venture
capitalist

Chemistry jobs | Chem-based jobs | Science-related jobs | Science-inspired jobs

If we look at the above chart, two promising areas for job growth are:

A. **Multidisciplinarity: expanding traditional disciplinary limits.**
It has often been said that chemistry is a central science. Traditionally, most people are trained in chemistry, biochemistry, or chemical engineering. As we contemplate additional job opportunities, the first extension can be made to areas such as biomedical, pharmaceuticals, advanced materials, new sources of energy, energy storage and transmission, bioengineering, nanotechnology, environment, ecology, and regulatory affairs. Further job extensions can be made to biology, computer science, food, agriculture, atmospheric science, geology, and physics. In yet a further extension, chemistry professionals may engage in careers that are loosely science-related or science-inspired, where their transferable skills can be applied.

Thus, chemistry professionals of the future need to be trained to be flexible, to learn different skills, and to use chemistry knowledge to tackle different jobs. In school, chemistry students will benefit from additional training or educational curricula that balance both depth and breadth of skills. Problem solving skills learned from studying chemistry are invaluable and applicable for a wide variety of careers.

B. **Globalization, expanding traditional geographic limitations.**
In this globalized environment, chemistry professionals should consider and be prepared for global job opportunities, e.g.,

- **Academia** - Postdocs in other countries, exchange program, sabbaticals, international teaching and research opportunities, international research collaboration, satellite labs

232

- **Industry** - International internship, overseas assignments within a company, global teams, clinical testing in other countries, sales support, import-export business
- **Small business** - Start-up, contract synthesis, contract manufacturing, international consulting business, global analytical and testing business
- **Others** - Teaching of English, editing of manuscripts, translation services, patent preparation

The main point is that for a job seeker in the future, there is a need to look for jobs beyond strictly bench research in core chemistry specialties and beyond geographical barriers in order to expand job options. For the chemistry enterprise, we need to facilitate this process, e.g., raising awareness of different options, providing training to increase eligibility for a broader spectrum of job options, and providing networking opportunities to connect with people in jobs outside of the laboratory.

6. *Supply/Demand of Scientists*

Even with job growth, we need a good balance of supply and demand. We cannot produce more workers than available jobs without leading to unemployment or underemployment. This holds true both in the United States and in the global setting.

A worrisome trend is the decreasing number of chemistry jobs from both NSF and Bureau of Labor Statistics (BLS) data in the past 20 years. In the past five years, the number of Ph.D.'s produced in the United States has increased from about 2,000 to 2,300 per year. It is clearly not a sustainable situation to have decreasing demand and increasing supply. A dialogue about the supply/demand balance is needed with U.S. stakeholders; some conversations can be held with other national chemical societies in the global context. ACS should take a leadership role in initiating these discussions.

In discussions with various ACS members across the country, there seems to be a wide range of viewpoints on the topic of supply and demand. From the point of view of most rank-and-file chemistry professionals, a large supply/demand ratio seems undesirable because it contributes to the current unemployment and underemployment situation, and tends to dampen salary increases. However, other opinions have also been expressed that the U.S. needs a large and talented STEM workforce in order to ensure that we have a steady supply of the best scientists in the world to compete in the future.

In view of the complexity of diverse views regarding supply and demand in the United States, it is recommended that a separate working group or task force be organized from interested ACS stakeholder committees and divisions to further examine and address this important topic in greater detail. As a result of this recommendation from the presidential Task Force, this action is now underway in discussions with the Committee on Economic and Professional Affairs taking a lead working with other interested stakeholder groups.

A. Employer versus Employee

In general, globalization tends to be beneficial for employers, but more uncertain for employees. For employers, globalization leads to ***greater access to*** world capital and markets, global tools and talent, and the potential to ***decrease cost*** by using these global resources. If an enterprise can maximize its use of resources, globalization can be very helpful.

However, for employees, globalization may lead to business consolidation, which can result in layoffs. Offshoring can lead to job loss in one sector of the economy and job creation in a different business sector. Furthermore, growth in new Ph.D.'s from emerging markets can increase competition in the global job market.

As an individual chemist, how can one take advantage of globalization? One option is not to function as an "employee" but as an "employer." The idea is to look for entrepreneurial opportunities created by globalization and start a business or partner with people on new ventures in the global market.

Nevertheless, not every chemist wants to be an entrepreneur. In that case, it pays to "think" like a business and manage one's career like a global business. For example, collaborate with others; increase productivity; and utilize global resources, tools, and talent.

B. The Global Perspective

Every outsourced American job is a job for a chemist somewhere. We need to support members worldwide in finding employment and empower all members (regardless of location) in seeking out productive careers.

C. Role of R&D Investment

Innovation is still the best engine for economic growth. Support for research and development is critical. Innovation can lead to new jobs, and support is needed for small businesses and entrepreneurs. ACS can be a catalyst for change in this area.

There is a need for increased government support for R&D to better prepare future scientists and to take advantage of the fresh talent pool to enhance U.S. global competitiveness. Enhanced funding for chemical education and increased support for small businesses and entrepreneurs will further support economic growth. For example, increased funding can be provided to existing federal initiatives, such as the Defense Advanced Research Projects Agency (DARPA), Small Business Innovative Research (SBIR), Small Business Technology Transfer (STTR), and other funds.

We can bring U.S. government, industry, and academia together to focus on supporting critical technology areas for the future. The partnership of academic research, the industrial research enterprise, and government support can facilitate intellectual property protection and commercialization, making them engines of growth for the 21st century chemistry enterprise.

A good example from history was Germany after World War II. Despite the problems after the war, Germany quickly recovered through effective partnership between the government, industry, and academia. The Industrial Technology Research Institute (ITRI) in Taiwan is another more recent successful example of government- academic-industrial cooperation.

D. **Help for Our Members**

The Task Force believes that it is critical to identify the ways globalization affects chemistry professionals, and to develop concrete recommendations and effective implementation plans for the professional growth and advancement of members.

The main benefits of globalization include

1. Increased job opportunities outside of the United States
2. Increased opportunities for partnerships between U.S.-based small and large businesses and global businesses based on mutual benefit
3. Facilitation of job creation in the United States through increased business activities
4. Enhanced cooperation between academic institutions
5. Greater options for sabbaticals and exchange programs
6. Ease of joint meetings of professional organizations connecting U.S. and worldwide research scientists

The full realization of these benefits requires **plans and actions:**

1. To publicize worldwide job opportunities, including exchange programs, internships, and temporary assignments
2. To provide information and training to prepare members for such opportunities
3. To increase interactions between national chemical societies in organizing international meetings, and to discuss issues that cut across borders relative to advocacy, education, public understanding, and balancing supply and demand for chemistry professionals

The challenges of globalization are numerous and we need to explore these challenges so that our members are prepared to meet them.

Many jobs are outsourced from one country to another. At the same time, chemistry students are being trained throughout the world, and globalization increases the migration of students and postdocs.

It will be useful to hold discussions with all stakeholders in order to bring a better equilibrium between supply and demand for chemical scientists. Job expansion can be sought in global, multidisciplinary, and non-research areas. A major emphasis should be placed on providing information on diverse career opportunities, global possibilities, entrepreneurship, and career management strategies and skills.

E. ACS and Sister Societies

As an organization, ACS should grow and seek opportunities from globalization. As described above, there will be different strategies for information services, meetings, and membership. As a part of the global chemistry enterprise, ACS needs to work with other national chemical societies. In particular, communicating chemistry's vital role to the public and policymakers, educational exchanges, developing a better understanding of supply and demand for chemists, and joint meetings and projects are primary areas to consider for collaboration.

8. *The Big Picture*

It is useful to summarize the aforementioned discussions in the following plot. The ACS is represented by the circle in the middle of the plot, both as a collection of members and as an organization. In the global environment, we see ACS as a premier professional society assuming an expansive role in chemical publications and meetings. To ensure U.S. competitiveness, the ACS needs to continue its advocacy for the chemistry enterprise, encourage increased federal R&D support, and promote entrepreneurship. Enhanced education and training need to be provided to ACS members to retool themselves, to provide more career options, and to expand global job opportunities. At the same time, the ACS and other national chemical societies need to work more closely together to foster collaboration and to address common problems such as enhanced public awareness of chemistry's vital role, chemical education, and supply/demand of chemists in order to benefit the global chemistry enterprise.

Summary –
Detailed ACS View

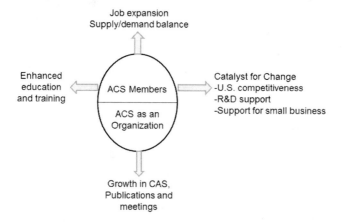

Another view is provided in the following diagram where the seven recommendations from the Task Force surround ACS in the center. ACS is an organization that can be seen playing at least three major roles:

A. ACS should serve as *a catalyst to inspire and initiate programs for chemistry communities* that advocate for the support of innovation and sustained investment in R&D, and promote entrepreneurship, supporting our strategic goal to "Communicate chemistry's vital role in addressing the world's challenges to the public and policymakers."

B. ACS should serve as *a facilitator,* collaborating with other national chemical societies to address common issues such as improved awareness with policymakers and the general public, education of future chemists, supply/demand of chemists, and global problems (such as the environment, sustainability, and maturing chemical industry) in order to benefit the global chemistry enterprise.

C. Finally, *ACS is an enabler,* leading in chemical information, publications, meetings, and education for the chemistry enterprise. Above all, the ACS is helping its members in job search and career development, supporting our strategic goal to "Empower an inclusive community of members with networks, opportunities, resources, and skills to thrive in the global economy."

237

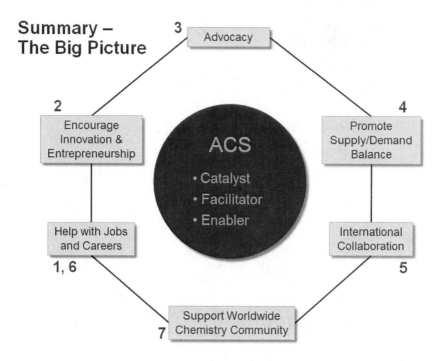

**Summary –
The Big Picture**

3 Advocacy

2 Encourage Innovation & Entrepreneurship

4 Promote Supply/Demand Balance

ACS
• Catalyst
• Facilitator
• Enabler

Help with Jobs and Careers
1, 6

International Collaboration
5

7 Support Worldwide Chemistry Community

VII. Conclusion

Although the job market does not appear to be improving substantially in the short-term, we firmly believe that ACS can play a critical role in helping members during these challenging times. Globalization is happening whether we like it or not. While globalization has its negative outcomes, it can also create a new set of opportunities. The key is to manage globalization in such a way to turn the *challenges into opportunities.* This is true not only for ACS, but also for individual ACS members.

The Presidential Task Force has planned programs for 2013 to showcase these opportunities at both the ACS National Meetings in New Orleans (April) and in Indianapolis (September). The proceedings of these Presidential Events will be documented in two ACS Symposium books, as well as streaming videos that will be accessible to chemistry communities everywhere. The opportunities that will be highlighted will cover a broad spectrum that applies to current students in the educational pipeline as well as mid-career professionals and members at a more advanced stage of their careers.

Working together, ACS, its leaders, volunteers, staff, and partners can substantially help ACS members thrive in the global chemistry enterprise. We truly believe in the presidential theme that we are all "Partners for Progress and Prosperity" where we can benefit by working together on common issues for mutual benefit.

VIII. Members of the Task Force

The Task Force, appointed by President-Elect Marinda Li Wu, is divided for operational purposes into two working groups: Globalization Opportunities and Jobs and Advocacy.

Globalization Opportunities

- Dr. H. N. Cheng, Co-Chair
- Mr. Jens Breffke
- Dr. Susan B. Butts
- Dr. Mukund S. Chorghade
- Dr. Peter K. Dorhout
- Dr. Cynthia A. Maryanoff
- Dr. Attila E. Pavlath
- Dr. Al Ribes
- Dr. Sonja Strah-Pleynet
- Dr. Shaomeng Wang
- Dr. Zi-Ling (Ben) Xue

Jobs and Advocacy

- Dr. Sadiq Shah, Co-Chair
- Dr. James Chao
- Dr. Pat N. Confalone
- Dr. Dan Eustace
- Dr. John Gavenonis
- Dr. Jennifer S. Laurence
- Dr. Zafra Lerman
- Ms. Connie J. Murphy
- Dr. Dorothy J. Phillips
- Dr. Joel I. Shulman
- Ms. Sharon Vercellotti

Dr. Wu leads the Task Force, along with the co-chairs. Dr. Robert H. Rich (ACS Director, Strategy Development) serves as staff liaison.

Biographies of Task Force Members

Mr. Jens Breffke is a graduate student at Penn State and member of the ACS Younger Chemists Committee (YCC), where he leads the International Presence (IP) working group. A native of Germany, he was chair of the German JungChemikerForum (JCF) of the German Chemical Society (GDCh) in 2006/07. In this function, he advocated and hosted together with his Hungarian colleague Csaba Janaky the founding meeting of the European Younger Chemist Network (EYCN) of the European Association for Chemical and Molecular Sciences

(EuCheMS) in Berlin in 2007. In his dual assignment in both the JCF as well as the EYCN, his continuous efforts was to promote international dialogues between younger scientists across their societies. The exchange program between the GDCh and the Northeastern Section of the ACS (NESACS) brought him as the German delegate to the ACS National Meeting in Boston in 2007. This experience and the continuing friendship with NESACS took a major part in his consideration of coming to the United States for graduate school. Jens organized a presidential event "Globalizing Education: Graduate School Opportunities in North America and Europe" which was a virtual symposium with most speakers presenting remotely using freeware. In 2010, he launched a pilot project of an exchange program between YCC and EYCN similar to the GDCh/NESACS program.

Jens received his Diplom Degree from the Humboldt-Universität zu Berlin, Germany in 01/09. Under his supervisor Nikolaus Ernsting, he designed and built a femtosecond pump-probe optical Kerr-shutter. He spent a research semester in the group of Manuel Mosquera González at the University of Santiago de Compostela, Spain, as a Leonardo-da-Vinci-II scholar of the German Academic Exchange Service (DAAD). Since 01/09 he studies ultrafast solvation dynamics in the group of Mark Maroncelli at the Pennsylvania State University as a graduate student.

Dr. Susan B. Butts is an independent consultant and an active member of the science and technology policy community following her 31 year career in the chemical industry and related organizations. Most recently, she served as the President of the Council for Chemical Research (CCR), a non-profit organization whose purpose is to benefit society by advancing research in chemistry, chemical engineering, and related disciplines through leadership collaboration across discipline, institution, and sector boundaries. Before joining CCR she worked for The Dow Chemical Company for three decades in various positions in the Research and Development organization. From 2001 through 2009, she served first as Director of External Technology then as Senior Director of External Science and Technology Programs. In that capacity, she was responsible for Dow's sponsored research programs at more than 150 universities, institutes, and national laboratories worldwide and also for Dow's contract research activities with U.S. and European government agencies. She also had responsibility for U.S. recruiting and hiring for R&D. She worked on issues related to science policy and government funding for research and development from Dow's office in Washington, DC. She is past president of the University-Industry Demonstration Partnership, an organization in the National Academies which works to strengthen research collaborations between universities and industry. She is also a member of the Council of the Government-University-Industry Research Roundtable in the National Academies and a member of the board of directors of the Alliance for Science and Technology Research in America. She is a Fellow of the American Association for the Advancement of Science, and a member of the ACS, Sigma Xi, and the Association of Women in Science.

Dr. Butts holds the degrees of B.S. in Chemistry from the University of Michigan and Ph.D. in chemistry from Northwestern University. Before joining the External Technology group Dr. Butts held several other positions at Dow including Senior Resource Leader for Atomic Spectroscopy and Inorganic Analysis within the Analytical Sciences Laboratory, Manager of Ph.D. Hiring and Placement, Safety and Regulatory Affairs Manager for Central Research, and Principal Investigator on various catalysis research projects in Central Research.

Dr. James Lee Chao received his B.S. and M.S. in Chemistry from the University of Illinois-Urbana in 1975 and 1976. He earned his Ph.D. in Physical Chemistry from the University of California-Berkeley in molecular spectroscopy in 1980.

Jim retired in 2009 from IBM Corporation after a 30 year career in the laboratory as a materials scientist and finishing as a strategist in business development for emerging technology commercialization. Jim has received industrial recognition for his many outside publications and chemical patents from his work. For many years, Jim also collaborated on developing novel methods for time-resolved infrared spectroscopy research as an adjunct Professor of Chemistry at Duke University. Most recently he served as the 2012 SERMACS Awards Program Chair for the ACS Southeast Region.

Jim has served the North Carolina local section in a number of capacities including Chair in 1991, alternate councilor in 1992, and councilor since 1993. While serving as chair-elect of the section, he reinstated the section's participation in Project SEED. In 1990, he was involved with local fund raising for the National ACS Campaign for Chemistry. From 1997-2001, he served on the International Activities committee. In 2002 he was appointed to the Committee on Patents and Related Matters and where he now serves as Chair. From 2008-11, he was subcommittee chair of National Awards with responsibility for preparing nominations for the ACS Board of Grants and Awards for the National Inventors Hall of Fame, the National Medal of Technology and Innovation, and the National Women's Hall of Fame.

In the area of outreach, Jim was responsible for introducing the North Carolina section to exhibiting at the NC State Fair in 1996 which has long been one of the section's largest participation events. Jim has served the section as the Undergraduate Scholarship Committee Chair from 1992-98 and has since been a member on the selection committee. In 1997, Jim was recognized by receiving the Marcus E. Hobbs Service Award from the section. Jim has been a Fellow of the American Institute of Chemists, a member of Sigma Xi, The Coblentz Society, and the New York Academy of Sciences.

Dr. H. N. Cheng is a Research Chemist at Southern Regional Research Center of the U.S. Department of Agriculture in New Orleans. He obtained his B.S. from UCLA and his Ph.D. from the University of Illinois at Urbana-Champaign. At USDA, he is working on several research projects involving product and process development of agricultural materials, particularly the conversion of agricultural waste and byproducts into valued added polymers. Previously, he worked for many years in industry in R&D, management, external technology, and new business development. He has developed many commercial products and has been associated with a wide range of R&D activities. Over the years, his research interest has included polymer chemistry, bio-catalysis, bio-based materials, food and nutrition, pulp and paper technology, and NMR spectroscopy. He has authored or co-authored 180 papers and 24 patent publications. He has organized or co-organized 22 symposia at ACS national meetings since 2003 and has co-edited 9 books. He is an ACS Fellow and an ACS Polymer Chemistry Division (POLY) Fellow.

At the ACS, he has been involved with numerous activities at the national, division, and local levels for many years. Currently, he is Chair of the International Activities Committee and 2013 Chair of the Louisiana local section. He also serves as a member of the 2013 Awards Equity Subcommittee for the Grants and Awards Board Committee, a member of Board of Trustees of the ACS Group Insurance Plan, co-chair of the POLY workshop committee, and a member of the POLY regional meetings committee.

Dr. Mukund S. Chorghade is President of Chorghade Enterprises and Chief Scientific Officer, THINQ Pharma/THINQ Discovery, AGN Biofuels and Empiriko. He is also an adjunct research professor at Northeastern University and has appointments at Harvard and MIT. He provides synthetic chemistry and development expertise to pharmaceutical and biopharmaceutical companies. He also provides consultations on collaborations with academic, government and industrial laboratories. He advises technology based companies on process re-engineering and project management of technology transfer; establishes strategic partnerships and conducts cGLP/cGMP compliance training and implementation in chemical laboratories. He oversees projects in medicinal chemistry, chemical route selection, manufacturing and formulation of bulk actives to finished dosage forms.

Dr. Chorghade earned his B. Sc. and M. Sc. degrees from the University of Poona, and a Ph. D. in organic chemistry at Georgetown University. He completed postdoctoral appointments at the University of Virginia and Harvard University, visiting scientist appointments at University of British Columbia, College de France/Universite' Louis Pasteur, Cambridge and Caltech and directed research groups at Dow Chemical, Abbott Laboratories, CytoMed and Genzyme. A recipient of three "Scientist of the Year Awards", he is an elected Fellow of the ACS, AAAS, and RSC and has been a featured speaker in several national and international symposia. He is active in ACS, AAAS, RSC, was NESACS and Brazosport Section Chair and serves on numerous professional Scientific Advisory Boards and Committees.

Dr. Pat N. Confalone is a recently retired Vice President of DuPont Global R&D, Crop Protection. Dr. Confalone has more than 42 years of experience in Pharmaceutical R&D at both Hoffmann-La Roche and DuPont Pharmaceuticals. He has directed both Medicinal Chemistry and Chemical Process R&D departments and led chemical development teams for the anti-hypertensive drug Cozaar and the HIV drug Sustiva, among others. Dr. Confalone has received a B.S. degree from MIT and Ph.D. from Harvard, under Nobel Laureate Professor Robert B. Woodward. He has more than 130 publications and 40 patents, has lectured extensively worldwide, and was elected a fellow of the AAAS in 2001.

Other honors include: Harvard Graduate Society Prize, 1968; Alpha Chi Sigma Award, 1967; Robert A. Welch Foundation Lecturer, 1988–89; Esther Humphrey Lecturer, 1990; and Samuel M. McElvain Industrial Speaker, 1982.

Industrial R&D experience includes DuPont, Global Research and Development, Crop Protection, vice president, 2003; Adaptive Therapeutics, Research and Development, vice president, 2003; Bristol-Myers Squibb, Process Research and Development, senior director, 2001–02; DuPont Pharmaceuticals, Chemical Process Research and Development, senior vice president, 1995–2001; and DuPont-Merck, Medicinal Chemistry, executive director, 1988–95.

His service in ACS National Offices include Board of Directors, District III, director, 2009–11; councilor ex officio, 2009–11; Committee on Budget and Finance, 2010–12, chair, 2011; Committee on Public Affairs and Public Relations, 2009–11; Committee on Chemistry and Public Affairs, 1995–2004, chair, 1997–98, committee associate, 1994, consultant, 2005–07; Green Chemistry Institute Governing Board, 2010–12; Presidential Task Force on Innovation in the Chemical Enterprise, 2010; and Task Force on National Institutes of Health, 1992–93. Other service in ACS offices: Member of ACS since 1970. *Division of Organic Chemistry:* chair, 1988–89; chair-elect, 1987–88; Executive Committee, 1985–90, chair, 1988; ACS Workshop on Chemistry, 1977.

Dr. Peter K. Dorhout is the dean of the college of Arts and Sciences at Kansas State University (KSU). Prior to coming to KSU, he served as interim provost at Colorado State University-Pueblo and as vice provost for graduate affairs and dean of the graduate school at Colorado State University in Fort Collins. His ACS leadership experiences include Board of Directors, District V director, chair of the committee on Professional and Member Relations, and member of the executive committee. He has been a member and chair of the Committee on Committees, Graduate Education Advisory Board, International Activities Committee, Colorado Section, and served on the Divisional Activities Committee and the Younger Chemists Committee.

Dr. Dorhout received his B. S. in chemistry from the University of Illinois, and his Ph. D. in chemistry from the University of Wisconsin and spent two years as a postdoctoral fellow at the Ames Laboratory at Iowa State University before joining the faculty of chemistry at Colorado State in 1991. His list of professional awards includes the Research Corporation Cottrell Scholar, Camille Dreyfus Teacher-Scholar, A. P. Sloan Fellow, and the ACS-Exxon Faculty. Award in Solid State Chemistry.

Dr. Dan Eustace has served members of several societies, local sections, and universities in sharing behaviors, emerging ideas, and best practices for managing careers. Most campuses, while well intentioned, are not staffed with career placement professionals who interview, work with and supervise technical "knowledge workers," like chemists, material scientists, biochemists and engineers. A unique focus is under-represented minority members.

At ACS, Dan serves as a career consultant and workshop presenter. In fun, informative and practical sessions he offers today's successful job search tactics and methods, tips for creating public relations documents (resumes, CVs, cover letters, etc.), and interviewing exercises that aim to improve job seekers' performance.

Dan retired from the staff of the film manufacturing division of Polaroid (2007; MCT) and staff scientist from ExxonMobil (1984) and now serves the Chemistry Department of the University of Connecticut as an adjunct. He has held staff and management positions in battery development, complex oil field chemical development, terrestrial solar cells, high tech film manufacture and environmental protection, industrial hygiene and chemical safety.

Dr. John Gavenonis is the Global Technology Manager - Renewable/ Sustainable Materials at DuPont Performance Polymers (DPP) in Wilmington, Delaware. In this position, he leads DPP's global R&D program to develop new engineering thermoplastic resins derived from renewable feedstocks. Prior to this role, John was part of the DPP industrial, consumer, and energy marketing group, where he expanded applications for DPP products in the global health care market segment. From December 2006 to April 2009, John was part of the Engineering Polymers technical service group with responsibility for nylon resins. He started at DuPont in November 2003 as a R&D Chemist in the business extensions group at DuPont Titanium Technologies.

John received an S.B. degree in chemistry from MIT, where he conducted undergraduate research with 2005 Nobel Laureate Professor Richard R. Schrock. He earned his Ph.D. in organometallic / inorganic chemistry in 2003 from the University of California, Berkeley under the direction of Professor T. Don Tilley. John is originally from Larksville, Pennsylvania.

John has been a member of the ACS since 1998, is a member of the Division of Inorganic Chemistry, is a former member of the California Section, and is currently Councilor and Government Affairs Committee Co-Chair for the Delaware Section. As Councilor, he is a member of the Local Section Activities Committee (LSAC) and is the LSAC liaison to the Committee on Chemistry and Public Affairs (CCPA). He was formerly Chair, Secretary, and NCW Chair of the Delaware Section. John is a member of the MIT Alumni Association Board of Directors, where he is developing a policy advocacy program for MIT alumni which is similar to the ACS Act4Chemistry Legislative Action Network (LAN).

Dr. Jennifer S. Laurence is Associate Professor in the Department of Pharmaceutical Chemistry at the University of Kansas School of Pharmacy. She received a Bachelors' degree from Miami University in 1994 and a Ph.D. in Chemistry from Purdue University in 2000, where she then pursued postdoctoral study in Structural Biology. Her research interests include protein and protein-conjugate stability, targeted delivery and controlled release of therapeutic metals, and mitigation of protein-based degradation of polymers.

Dr. Laurence was awarded the 2011 Louise Byrd Graduate Educator Award for extraordinary devotion to graduate students and their education and for distinguished scholarship. She was awarded Teacher of the Year by the American Association of Colleges of Pharmacy (AACP) in 2006. In 2012, she launched a partnership with Nairobi University for a distance Master's degree program.

Dr. Laurence serves on the American Chemical Society Committee on Science as executive chair of Public Policy and Communication. She is on the Editorial Advisory Board for the *Journal of Pharmaceutical Sciences* and was Guest Editor for a 2012 special issue of *Molecular Pharmaceutics* on "Advances in Biophysical and Bioanalytical Protein Characterization". She has served on NIH and NSF grant review panels covering grant applications related to biotechnologies, protein characterization, analytical tool development, and cancer detection, and is a consultant to major pharmaceutical and biotechnology companies, start-ups, and contract research organizations.

Her approach to research is multidisciplinary and translational, and she has collaborated with many partners, including industry, clinicians, business consultants, intellectual property experts and other academics. In 2011, Dr. Laurence received a Wallace H. Coulter Translational Research Award. She founded Echogen Inc. and raised significant investment capital to develop her patented metal abstraction peptide (MAP) technology.

Dr. Zafra Lerman is the President of MIMSAD (Methods Integrating Music, Science, Art and Dance). Her Ph.D. is from the Weizmann Institute of Science, Israel, and she conducted research at Cornell and Northwestern Universities, and the ETH, Zurich, Switzerland. She developed an innovative approach of teaching science at all levels using the arts and cultural backgrounds, which received international recognition, and she has lectured around the world. For 25 years, she has chaired the sub-Committee on Scientific Freedom and Human Rights for the American Chemical Society (ACS). At great risk to her safety, she was successful in preventing executions, releasing prisoners of conscience from jail and bringing dissidents to freedom. Since 2001, she has been using chemistry as a bridge to peace in the Middle East. She is the president of the "Malta Conferences Foundation" which brings together scientists from 15 Middle East countries with six Nobel laureates to work on solving regional problems, establishing cross-border collaborations, and forging relationships that bridge chasms of distrust and intolerance.

Dr. Lerman has received 38 national and international awards for her work such as the Presidential Award from President Clinton (1999); the World Cultural Council's World Award for Education in Johannesburg, South Africa (2000, the first international award in the new democratic South Africa); the ACS Parsons Award for outstanding public service to society through chemistry (2003); The Royal Society of Chemistry, England, Nyholm Education Award (2005); New York Academy of Sciences Pagels Human Rights Award (2005); George Brown Award for International Scientific Cooperation from CRDF Global(2007); the ACS Pimentel Award for excellence in chemical education (2010); and the Peace Award from the International Center for Innovation in Education (2010). She serves as Vice Chair for Chemistry of the Board of the Committee of Concerned Scientists (CCS); and is a Fellow of the ACS, the American Association for the Advancement of Science (AAAS), the Royal Society of Chemistry (RSC), and the American Institute of Chemists (AIC).

Dr. Cynthia A. Maryanoff is a Distinguished Research Fellow in the Product and Process Scientific Solutions affiliate of Johnson & Johnson. Cyndie began her 30+ year career with J&J in 1981 when she joined McNeil Pharmaceutical as Section Head in Chemical Development advancing to global head of the Chemical and Pharmaceutical Development of Drug Evaluation for J&J PRD. There she worked to rapidly move new molecular entities from small-scale chemical synthesis to large-scale, from development of oral formulation and toxicology studies to first-in-human and proof-of-principle studies in record times relative to industry standards. Cyndie is very active in the scientific community both locally and nationally, especially within the ACS Division of Organic Chemistry where she currently serves on the Executive Committee (1988-present). An inaugural ACS Fellow, Cyndie is also the recipient of numerous national, local and internal awards which encompass both scientific and managerial achievements most notable: The ACS Earle B. Barnes Award for Leadership in Chemical Research Management; The ACS Garvan – Olin Medal; and the ACS Henry F. Whalen, Jr. Award for Business Development. Cyndie is a Fellow of the American Association for the Advancement of Science and is a member of the Advisory Council for the College of Arts and Sciences at Drexel University. She holds a B.S. degree from Drexel University, a Ph. D. in chemistry from Princeton University with Prof. Kurt Mislow, and postdoctoral training with Prof. Edward C Taylor.

Cyndie's ACS service includes: ACS Development Advisory Board, 2011-2014; ACS Task Force on Multidisciplinarity of Chemistry, 2004-2005; Advisory Board of Journal of Organic Chemistry, 2000-2004; Committee on Science, 1992-1995; ACS Books Advisory Board, 1994-1997; Advisory Board; Chemical & Engineering News, 1990-1992; Advisory Board ACS PRF, 1986-1989; *Division of Organic Chemistry:* Executive Committee (1988-present); Chair, 1997; Councilor or Alternate Councilor, 1992-2003, 2010-2012; Student Travel Awards Committee Chair, 1990–present; Chaired and organized 22 ACS Award Symposia; *Division of Medicinal Chemistry:* Long Range Planning Committee, 1999-2003; Study Section NIH Grants, 1988-1992; NIH Medicinal Chemistry Small Business Innovation Research Program Review, 1993; Women Chemist Committee: Committee Associate, 1995; Search Committee for Editor of JOC (1999); Chair, Search Committee for Editor of Accounts of Chemical Research (1995).

Ms. Connie J. Murphy retired in 2008 after working for more than 28 years at The Dow Chemical Company in Midland, MI. She spent the first 25 years of her career in R&D in roles including research technologist, team leader, group leader and project portfolio manager. She holds five patents in monomer/polymer synthesis and processing. The last three years of her career at Dow she was a supervisor of IT professionals in information systems.

Connie has been a member of the ACS since 1992. Since joining ACS, she served as a member of the Committee on Committees, the Committee on Membership Affairs, the Committee on Technician Affairs and the Chemical Technology Program Approval Service. She is the 2011 chair of the ACS Committee on Chemistry and Public Affairs and the 2011 chair-elect of the

Industrial and Engineering Chemistry Division. She served in the leadership of the Division of Chemical Technicians for several years, serving as chair-elect, chair, past chair, councilor, webmaster, membership chair and member-at-large. In the Midland Section of the ACS, she served as a director and membership chair for several years and is currently the government affairs committee chair and alternate councilor. She facilitates two of the workshops ("Leading Change" and "Coaching and Feedback") in the ACS Leadership Development System. Connie was honored by her selection as an ACS Fellow in 2011.

She received the Outstanding Service Award from the Midland Section of the ACS in 2008, Distinguished Service Award from the ACS Division of Chemical Technicians in 2006, Outstanding Chemical Technician Award from the Midland Section of the ACS in 1997, and Outstanding Technologist Achievement Award from the Dow Central R&D Scientists Organization in 1994. She also received seven Special Recognition Awards from The Dow Chemical Company between 1988 and 2002 for technical and professional contributions.

Dr. Attila E. Pavlath is a Senior Emeritus Research Chemist at the Western Regional Research Center of the U.S. Department of Agriculture in Albany, California, conducting research on the utilization of agricultural products as chemical resources. He finished his studies in Hungary, where he received his doctoral degree in chemistry from the Hungarian Academy of Sciences. He was an Assistant Professor at the Technical University of Budapest and when he left Hungary in 1956. He was first at McGill University in Montreal, Canada until 1958, and then he moved to California when he joined the Western Research Center of Stauffer Chemical Company in Richmond, California as a Senior Group Leader. Since 1967 he is with the U.S. Department of Agriculture in Albany, leading various research projects. His research activities since 1951 earned him international recognition as an expert in a wide variety of scientific areas. He has been involved in fluorine chemistry, glow discharge chemistry, textile chemistry, energy research and agricultural chemistry. He is an internationally known expert on these fields with more than 130 scientific publications and 25 patents. He has lectured throughout the world and speaks four languages.

His activities in the ACS dated back to 1969. He chaired the California Section (three times) and the Division of Professional Relations. He chaired many committees at National and Local level. He was elected three times to the ACS Board of Directors before he was elected as ACS President for 2001. He is still involved in various ACS activities.

Dr. Dorothy J. Phillips was Director of Strategic Marketing at Waters Corporation prior to her retirement in April 2013. She holds a B.A. degree from Vanderbilt University and a Ph.D. from the University of Cincinnati. Dorothy began her thirty-nine year industrial career at Dow Chemical Company where she spent nine years and received three patents in the area of animal growth promoters. She joined Waters Corporation in R&D in 1984. While in R&D she was instrumental in the development of Waters' products such as Accell® stationary phases, Symmetry® columns and Oasis® sorbent. She received the Waters Manager's Award for Innovation in 1987 and 1988. Dorothy has published and/or presented over 70 papers in the field of analytical chemistry with HPLC focus and made presentations worldwide at scientific meetings and at Waters customers' venues. As Director, Strategic Marketing, she was responsible for identifying and assessing new technologies, business and product opportunities to meet global separation needs in pharmaceutical, biopharmaceutical and food safety markets. Dorothy traveled extensively in Europe, Asia and Japan; prior to the Olympics in China in 2008, she worked closely with her local colleagues and the government agencies to understand analytical testing required for food and water safety. Dorothy was the first ever recipient of the Waters Leadership Award for outstanding contributions to Waters and the Waters Community in 2008.

She has received the following honors: the American Chemical Society Fellow, Class of 2010; Distinguished Chemist Award, The New England Institute of Chemists (NEIC), Division of the American Institute of Chemists, 2011; Shirley B. Radding Award, Santa Clara Valley Section, ACS, 2008; ACS Northeastern Section Henry A. Hill Award, 2006; Nashville Section ACS, Salute to Excellence Award, 2004; Vanderbilt University, Dr. Dorothy Wingfield Phillips Award for Leadership, 2007; Unsung Heroine Award, Vanderbilt University, 2006; Distinguished Alumni, University of Cincinnati, McMickens College of Arts and Sciences, 1994 and Center for Women Studies, 1993.

Dorothy served in ACS national offices as a member of the Council Policy Committee (2008-13), the Committee on Committees (2001-06), and the Committee on Divisional Activities (2007-08). She also served in the Division of Analytical Chemistry as Chair (2009-10), Program Chair (2008-09) and Chair-Elect (2007-08). Her service in NESACS offices include Councilor (1995-2015), Chair (1993), Chair-Elect and Program Chair (1992), Awards Committee, Chair (2009-13), Fundraising Committee, Chair (2004-08), Project SEED Committee, Chair (1994-95), and Centennial Celebration, Co-chair (1998). She was recently elected to the NESAC office of Trustee for 2014-2016.

Dr. Al Ribes is a Senior Lean Six Sigma Consultant for Dow Benelux. Al applies the scientific method to solving business problems, achieving improvements, and reducing waste. Al has worked for over twenty years in industry at two companies and in four countries: Rio Tinto Mining Co (Spain), and Dow Chemical in the United States (Louisiana, Texas), Argentina, and the Netherlands. Al earned a PhD from SUNY Buffalo in Electro-analytical chemistry. Upon joining Dow, Al developed expertise in long chain branching and molecular weight characterization using HTGPC. He has also worked on international technology transfer. In 2001, Al left the bench and moved to Six Sigma; he certified as a Master Black Belt, which is the highest technical leadership role in the field of Six Sigma. Al has served as treasurer and newsletter editor of the ACS Analytical Division, has served in the ACS Committee on Community Affairs, and currently chairs the ACS Committee on Minority Affairs. Al lives in the Netherlands.

Dr. Robert H. Rich is the Director, Strategy Development for the American Chemical Society. In this capacity, he supports the Board of Directors, management, and leadership in strategic planning, environmental scanning, scenario planning, and consideration of major strategic issues such as worldwide strategy and sustainability. He holds a Bachelors' degree from the Massachusetts Institute of Technology, a Masters from Harvard University and a Ph.D. from the University of California at Berkeley, all in chemistry, and is a Certified Association Executive. Previous roles at the ACS have included manager of professional development services, program officer at the ACS Petroleum Research Fund, and divisional web manager. Before coming to the ACS, Bob worked as a research fellow at the National Institutes of Health and on the staff of the American Association for the Advancement of Science (AAAS) in its Directorate for Science and Policy Programs. In these roles, he has supported and facilitated many conferences and strategic discussions.

Dr. Sadiq Shah is the Vice Provost for Research at the University of Texas - Pan American with responsibilities for managing, directing and growing the research and scholarship activities as well as technology transfer efforts on campus. Prior to this Dr. Shah served as the Associate Vice President for Research at the California State University Channel Islands, and also served as the Associate Vice President for Research and Economic Development at Western Kentucky University (WKU). He was responsible for the Offices of Sponsored Programs, Technology Transfer, and Economic Development; the WKU's Center for Research and Development; the Small Business Development Center; and served as the Chief Executive Officer and Chair of the Board of Directors for the WKU Research Foundation. Prior to that, he served as the founding Director of the Western Illinois Entrepreneurship Center at Western Illinois University. During his tenure at Western Illinois University, Dr. Shah in his role was responsible for the Office of Technology Transfer, three Entrepreneurship Centers in West Central Illinois, the Executive Studies Center in Macomb, the Small Business Development Center, the Center for The Applications of Information Technologies, the Illinois Manufacturing Extension Center for West Central Illinois, and the Procurement Assistance Centers. Dr. Shah's efforts helped facilitate the start-up of 23 companies in West Central Illinois.

After receiving his doctorate in chemistry from Washington University, St. Louis, in 1986, Dr. Shah joined Petrolite Corporation as a Senior Research Chemist with responsibilities for New Technology Development. In 1991 he joined Calgon Vestal Laboratories, A subsidiary of Merck Pharmaceuticals, in St. Louis as a Group Leader for Product and Technology Development. The ownership of the company evolved from Merck to Bristol-Myers Squibb and then STERIS Corporation. Dr. Shah's responsibilities with the company evolved in management of the R&D efforts in Infection Control, Wound Management, Decontamination, Product and Process Development and the Engineering group with responsibilities for product delivery technologies at different stages of his tenure with the company. Dr. Shah has been responsible for guiding the development of 20 new products from concept to launch and seven technology platforms. Dr. Shah has 15 patents. Dr. Shah has edited a book, written book chapters, and published 30 research articles and other articles related to technology transfer.

Dr. Joel I. Shulman is an Adjunct Professor of Chemistry at the University of Cincinnati. After obtaining a B.S. degree from The George Washington University in 1965, he received his Ph.D. in organic chemistry in 1970 from Harvard University. In 1970, he joined the research staff of the Procter & Gamble Company (P&G). During his 31-year career at P&G, he managed projects ranging from drug discovery to the manufacture and commercialization of decaffeinated instant coffee brands to developing ingredients for the first 2-in-1 shampoo. From 1996 to 2001, he was Manager of External Relations and Associate Director of Corporate Research at P&G, with responsibility for bringing new technical capabilities into the company. Included in his department were doctoral recruiting, university relations, external research programs, interactions with government laboratories, and technology acquisition from Russia and China.

Upon retiring from P&G in 2001, Joel joined the faculty at the University of Cincinnati, where he teaches undergraduate organic chemistry and a course called "Life after Graduate School." He developed this latter course into a two-day workshop presented by the ACS entitled "Preparing for Life after Graduate School." Joel serves the ACS as a Career Consultant, a consultant to the Graduate and Postdoctoral Scholar Office, Chair of the Graduate Education Advisory Board, and a member of the Committee on Professional Training. He is a Fellow of the ACS.

Dr. Sonja Strah-Pleynet has 10 years of experience in the biotechnology / pharmaceutical industry and has been an active ACS member since 1998. She has held various leadership positions on the Executive Board of the San Diego Local Section, including Councilor and Chair of the Government Affairs Committee. At the national level, she served on the Awards Committee, Committee on Economic and Professional Affairs (CEPA) and has been a CEPA liaison to the Committee on Minority Affairs and International Activities Committee. Under her leadership, San Diego Local Section won a ChemLuminary Award - ACS President's Award for Local Section Government Affairs in 2010.

Dr. Strah-Pleynet received a Ph.D. in Organic Chemistry from the University of Ljubljana, Slovenia and moved to the United States in 1997, after being awarded a postdoctoral fellowship with Prof. Alan Katritzky at the University of Florida. She started her professional career in 1999 at Arena Pharmaceuticals where she held positions with increased responsibilities for ten years before moving to Boston in 2010. She made key contributions to various multidisciplinary programs in discovery and development of novel therapeutics for CNS, cardiovascular, inflammatory and metabolic diseases, including two clinical candidates, APD125 for insomnia and APD179 for cardiovascular disease. She is a co-inventor on 10 US patent applications, co-author of more than 30 publications and has presented at several national and international conferences. She has often shared her experience in pharmaceutical industry with students, teachers and younger chemists through classroom visits and career development symposia.

As the Chair of her state's ACS Government Affairs Committee for four years, she organized and led federal legislative district office and Capitol Hill visits to engage legislators and advocate on issues of importance to the Society and its members, such as science research funding, STEM education, innovation, green chemistry and sustainability. In this role, she initiated and developed collaborations and partnerships between industry and academia, ACS and other scientific and professional organizations.

Ms. Sharon Vercellotti founded V-LABS, INC., a business specializing in carbohydrates, with the vision of making available both products and services to researchers in biotechnology, biochemistry, and in the developing glycobiology with its potentially revolutionary approaches in solving biomedical problems. V-LABS, INC., serves customers in the United States and internationally.

She is an active ACS member and member of CEPA. She has served the Division of Small Chemical Businesses as Councilor (present), Alternate Councilor, Chair, Chair Elect and Secretary. She developed the SCHB web page and was web master for several years. Currently she is editor of the Division's newsletter, Small Chem Biz. She is a member of the Division of Carbohydrate Chemistry. Vercellotti has organized symposia at both national and regional ACS meetings for SCHB and the Biotechnology Secretariat. She has served on the Committee on Technician Affairs, 2002-2009. She was selected Top DOG by the ACS Divisional Officers Group in 2003. She was selected Fellow of the American Chemical Society in 2011.

Vercellotti served on the NSF National Visiting Committee for the Project to Support Chemistry-based Technical Education under the ACS from 2000-2003. She was an NSF Advisory Board Member of the Industrial Innovation Interface (SBIR program) from 1989-1994, and vice-president of the Louisiana Alliance for Biotechnology from 1997-2001.

Dr. Shaomeng Wang received his B.S. in Chemistry from Peking University in 1986 and his Ph.D. in Chemistry from Case Western Reserve University in 1992. Dr. Wang did his postdoctoral training in drug design at the National Cancer Institute, NIH between1992-1996. Dr. Wang was Assistant Professor at Georgetown University from 1996-2000 and Associate Professor from 2000-2001. Dr. Wang joined the faculty at the University of Michigan Medical School as a tenured Associate Professor in 2001 and was promoted to Professor in 2006. Dr. Wang was named the Warner-Lambert/Parke-Davis Professor in Medicine in 2007. Dr. Wang serves as the Co-Director of the Molecular Therapeutics Program at the University of Michigan Comprehensive Cancer Center and is the Director of the Cancer Drug Discovery Program at the University of Michigan.

Dr. Wang is the Editor-in-Chief for ACS Journal of Medicinal Chemistry, a premier international journal in medicinal chemistry and drug discovery and serves on the editorial board for several international journals.

Dr. Wang has published more than 200 papers in peer-reviewed scientific journals and 100+ meeting abstracts, and is an inventor on more than 40 patents and patent applications. In addition to his academic role, Dr Wang is a co-founder of Ascenta Therapeutics and Ascentage Pharma, which were established to develop innovative anticancer medicines with technologies licensed mainly from the University of Michigan.

Dr. Marinda Li Wu received a B.S. *cum laude with Distinction in Chemistry* from The Ohio State University in 1971 and a Ph.D. in Inorganic Chemistry from the University of Illinois in 1976. With over thirty years of experience working in the chemical industry, she enjoyed many years working for Dow Chemical R&D as well as Dow Plastics Marketing forging partnerships between industry, education, government and communities. Dr. Wu also has entrepreneurial experience with various small chemical companies and startups including *"Science is Fun!"* which she founded to engage young students in the excitement of science and enhance public awareness of the importance of supporting and improving science education.

As an ACS member for over forty years, Dr. Wu has served in many leadership roles at both the local and national levels for the American Chemical Society. Dr. Wu was elected to the ACS Board of Directors in 2006, and served as a Director-at-Large until 2011, when she was elected to the Presidential succession of the American Chemical Society. As ACS President-Elect for 2012, she was invited to give plenary lectures worldwide and made an honorary member of the Romanian Chemical Society and Polish Chemical Society. She serves as ACS President in 2013 and Immediate Past President in 2014.

Dr. Wu serves on the University of Illinois Chemistry Alumni Advisory Board, the International Advisory Board for the 45th IUPAC World Chemistry Congress 2015, the ACRICE-1 (1st African Conference on Research in Chemistry Education) International Advisory Board, and the Board of Directors for the Chinese-American Chemical Society. She holds 7 U.S. Patents and has published a polymer textbook chapter and numerous articles in a variety of journals and magazines over the years.

Dr. Zi-Ling (Ben) Xue received his BS degree in 1982 from Nanjing University-Nanjing College of Pharmacy, China, and Ph.D. degree from the University of California at Los Angeles in 1999. He then conducted postdoctoral studies at Indiana University in 1990-1992, and accepted a position as Assistant Professor at the University of Tennessee in 1992. He is now a Paul and Wilma Ziegler Professor of Chemistry with research in inorganic and analytical chemistry. Dr. Xue has received several awards and honors including a National Science Foundation (NSF) Young Investigator Award, NSF Special Creativity Award, Camille Dreyfus Teacher-Scholar Award, Changjiang Lecture Professor, and Distinguished Oversea Young Scholar Award (Chinese Natural Science Foundation). He is a fellow of the American Association for the Advancement of Science.

Dr. Xue is Membership Chair, ACS Division of Inorganic Chemistry. He has served in the ACS Multidisciplinary Program Planning Group (MPPG) and its executive committee to select themes for ACS National Meetings. He initiated and helped the ACS Office of International Activities to host the first visit of a Chinese chemistry dean's delegation in 2008. He also initiated and organized the China-US Chemistry Deans/Chairs Forum in Beijing in 2009.

References and Notes

1. For more on ACS Career Management and Development resources, see www.acs.org/careers. For more on ACS advocacy efforts, see www.acs.org/policy. The ACS maintains an active Legislative Action Network, which provides members the opportunity to influence policymaking while key issues are being legislated. For more on ACS International Activities, see www.acs.org/international.
2. This recommendation involves leadership from the Committee on Economic and Professional Affairs.
3. Could tie in to the ACS Network "Your Profession" section, which we could populate and members could edit as well.
4. This will be coordinated by the Professional Education group. We are developing this kind of online tool for accessing information relevant to differing member segments.
5. ACS is exploring brief video tutorials, to simply illustrate the steps in job searching. We hope to produce some examples in 2013. We did establish an "ACS Careers" channel on YouTube. Also, the ACS Network can support videos.
6. This recommendation involves leadership from the Entrepreneurial Initiative Advisory Board and other relevant committees and stakeholders.
7. ACS is currently planning entrepreneurial showcase events to take place in December 2012. The Entrepreneurial Initiative will continue through 2013.
8. This will be coordinated by the ACS Graduate Education Community and other collaborative partners.
9. ACS currently offers Preparing for Life after Graduate School workshops, which are being evaluated and refined.
10. On the ACS Network, under "Your Profession," there is a section for chemical entrepreneurs. The Chemical Entrepreneurs Council also has a section there. We are building a nexus for everything related to supporting chemical entrepreneurs.
11. This involves leadership by the ACS Board of Directors, the Graduate Education Advisory Board, and other leaders in the ACS graduate education community. ACS has piloted workshops for faculty, and the Presidential Graduate Education Task Force has proposed several new approaches.
12. This recommendation involves leadership from the Committee on Public Affairs and Public Relations, the Committee on Chemistry and Public Affairs, and other relevant committees.
13. The ACS Chemistry Ambassadors Program is working to do this. It would be helpful to document or record these resources and create a webcast.
14. ACS currently has a wide network of Local Section Government Affairs Committees across the United States.
15. Involves the leadership of the Committee on Corporation Associates
16. See ACS position statement on "A Competitive U.S. Business Climate: Innovation, Chemistry, and Jobs" at www.acs.org/policy. With proper training, ACS members can play an important role to advocate for legislation

that will make domestic business more competitive so as to improve the jobs climate in United States.

17. This discussion should be led by the Committee on Economic and Professional Affairs and involve the Committee on Professional Training, Graduate Education Advisory Board, Younger Chemists Committee, and other interested parties.

18. This recommendation involves leadership from the International Activities Committee.

19. These could also involve participation of ACS Technical Divisions and other national chemical societies.

20. This recommendation involves leadership from the Committee on Economic and Professional Affairs and the International Activities Committee.

21. The Committee on Professional Training has initiated some discussions with Europe, and this could be expanded to include other regions such as Asia and Latin America.

22. This recommendation involves leadership from the International Activities Committee.

23. Please see http://strategy.acs.org for more information on the *ACS Strategic Plan for 2013 and Beyond.*

24. Source: *ACS Office of Research and Member Insights*, 2012.

Editors' Biographies

H. N. Cheng

H. N. Cheng (Ph.D., University of Illinois) is currently a research chemist at Southern Regional Research Center of the U.S. Department of Agriculture in New Orleans, where he works on projects involving improved utilization of commodity agricultural materials, green chemistry, and polymer reactions. Prior to 2009, he was with Hercules Incorporated, where he was involved (at various times) with new product development, team and project leadership, new business evaluation, pioneering research, and supervision of analytical research. Over the years, his research interests have included NMR spectroscopy, polymer characterization, biocatalysis and enzymatic reactions, functional foods, and pulp and paper technology. He is an ACS Fellow and a POLY Fellow and has authored or co-authored 180 papers, 24 patent publications, co-edited nine books, and organized or co-organized 22 symposia at national ACS meetings since 2003.

Sadiq Shah

Sadiq Shah (Ph.D., Washington University, St. Louis) is the Vice Provost for Research at the University of Texas-Pan American with responsibilities for managing, directing, and growing the research, scholarship, and creative activities as well as technology-transfer efforts on campus. Earlier, Dr. Shah served successively as Associate Vice President for Research at California State University Channel Islands, Associate Vice President for Research and Economic Development at Western Kentucky University, Director of the Western Illinois Entrepreneurship Center and the Office of Technology Transfer at Western Illinois University, Manager, Product & Technology Development for Health Care markets at STERIS, BMS and Merck, and senior research chemist at Petrolite Corporation. He has been responsible for guiding the development of 20 new products from concept to launch and seven technology platforms. He has 15 patents, has edited a book, written book chapters, and published 30 research articles and other articles related to technology transfer.

Marinda Li Wu

Marinda Li Wu (Ph.D., University of Illinois) is the 2013 President of the American Chemical Society (ACS). She has over 30 years of industrial experience at Dow Chemical R&D and Dow Plastics Marketing, and additional entrepreneurial experience with various small chemical companies and startups including "Science is Fun!" which she founded to engage young students in science and enhance public awareness. She has served in many leadership roles at local and national ACS levels. She was elected to the ACS Board of Directors in 2006 and served as Director-at-Large until 2011. In 2011, she was elected to the Presidential succession of the American Chemical Society, where she brought fresh ideas, boundless energy, and enthusiasm for science to chemistry communities around the world. She holds seven U.S. patents and has published a polymer textbook chapter and numerous articles in a variety of journals and magazines over the years.

Indexes

Author Index

Subject Index